AF615080

American Airlines, Inc.

$$\sigma_k = \sum_{n=1}^{\infty} \frac{(A_n)_k}{M_n \omega_n} \int_0^t Q_n(t') \sin \omega_n(t - t')dt'$$

Structural failure of an airplane wing caused by a severe impact at landing, and the equation for the stresses in an elastic structure caused by an impact (see Chapter 7).

AN INTRODUCTION TO THE DYNAMICS OF AIRPLANES

H. NORMAN ABRAMSON

MANAGER, ENGINEERING ANALYSIS
SOUTHWEST RESEARCH INSTITUTE

DOVER PUBLICATIONS, INC.
NEW YORK

Published in Canada by General Publishing Company, Ltd., 30 Lesmill Road, Don Mills, Toronto, Ontario.

Published in the United Kingdom by Constable and Company, Ltd., 10 Orange Street, London WC 2.

This Dover edition, first published in 1971, is an unabridged and corrected republication of the work originally published by the Ronald Press Company in 1958. Appendix C, Solutions to Problems was originally published separately in 1959 by the Ronald Press Company as a supplement to this book.

International Standard Book Number: 0-486-60508-6
Library of Congress Catalog Card Number: 73-153056

Manufactured in the United States of America
Dover Publications, Inc.
180 Varick Street
New York, N. Y. 10014

PREFACE

The very rapid growth of aeronautical science in recent years has placed increasing importance upon the dynamical problems involved in aircraft design, construction, and operation. This group of problems, of great practical importance and considerable complexity, may be gathered together under the single topic of "dynamics of airplanes."

A competent worker in the field of airplane and missile dynamics must be able to integrate the knowledge, methods, and techniques of many different fields. He must possess a rather complete knowledge of structural analysis, aerodynamics, vibration theory, and experimental procedures. He must have a better-than-average working knowledge of mathematics, at least through differential equations and preferably through the whole gamut of topics known collectively as "advanced calculus." Finally, he should have a knowledge of the methods of automatic machine computation and the analysis of electrical circuits. Since the computations involved in many dynamics problems are so intricate and tedious, ever-increasing use is being made of electrical analogue methods and automatic computing machines.

Unfortunately, there has been, and still is, a shortage of scientifically trained personnel to cope with such problems. Accordingly, many universities are now introducing courses which are designed to provide the forthcoming graduate in aeronautical engineering with some knowledge concerning this rapidly expanding branch of the science.

The purpose of this book is to present the basic principles and ideas concerning airplane dynamics problems which should be included in the "tool kit" of today's aeronautical engineering graduate. The material presented here purposely avoids concentration on the details of any particular analysis, but rather tries to present the general picture of airplane dynamics problems. This volume is not intended as a reference text for dynamics specialists; instead, it is written on a level for senior college students in aeronautical engineering who have had only the usual second- or third-year course in elementary dynamics and an introductory course in differential equations. The student who has digested the contents of this volume should have little difficulty in mastering the details of any given procedure or technique and in progressing to more advanced studies. This book may also be useful in an introductory graduate course if sufficient additional material is provided by the instructor and adequate use is made of the references listed at the end of each chapter. It is the author's hope, additionally, that practicing engineers in the aeronautical industry who are encountering dynamical problems for the first time may find this volume a useful introduction to more advanced books and to the rapidly increasing number of technical papers in the field.

Most of the problems encountered in the study of airplane and missile dynamics have the common feature that they are principally vibration problems. It may be that in order to write the differential equations of a given problem it is necessary to employ a structural analysis to determine some terms, and an aerodynamic analysis to determine others. The equations may require the use of automatic computing machines to arrive at their solutions. Experimental studies may be required to check the accuracy of some assumptions and computations. But the problem is still basically one of vibration, and therefore the student will perhaps not be too surprised to find that the first third of the book deals with vibration theory. The remaining two-thirds of the book do indeed treat the dynamics of airplanes.

Three further points concerning this book are perhaps worthy of special mention. First, the almost universal use of matrix algebra in the literature concerning airplane dynamics has demanded its inclusion in this text, and therefore a brief treatment is given in Chapter 3. Second, because in most undergraduate aeronautical engineering curriculums only passing mention is made of the subject of flight stability of aircraft, Chapter 8 is devoted to an introductory treatment of this important problem. Third, the problems have been rather carefully selected. In most instances, they constitute amplifications or extensions of the text and should therefore be regarded as much a part of the theoretical development as the text itself.

Most of the material presented in this book has been used in lecture form as a one-semester course of three hours in the Department of Aeronautical Engineering of the Agricultural and Mechanical College of Texas, and in a series of lectures at Southwest Research Institute. It is emphasized, however, that the content of this volume exceeds that which can be treated adequately in a three-hour course. This will allow the instructor some degree of choice in the material to be presented. If, for instance, the students have already had a course in vibration theory, he may choose to begin with Chapter 3, thus allowing additional time for the remaining chapters. Chapters 7 and 8 are independent and hence either may be studied immediately following Chapter 6, or omitted entirely, according to the demands of the curriculum or the wishes of the instructor. Chapter 9 contains a general discussion of certain miscellaneous and advanced topics as well as a general survey of dynamic aeroelasticity and may therefore be included in the classroom discussion or be left for independent study by the student.

No book can be written without the advice, counsel, and encouragement of others. This one is certainly no exception, and I therefore gratefully acknowledge this help from Professor E. A. Ripperger of the University of Texas, Professor Glenn Murphy of Iowa State College, and from T. B. Epperson and Martin Goland of Southwest Research Institute. My thanks, also, to my wife for her patience and constant encouragement and to Mrs. E. F. Gee for her careful and efficient typing of the manuscript.

H. Norman Abramson

San Antonio, Texas
January, 1958

CONTENTS

AN INTRODUCTION TO THE DYNAMICS OF AIRPLANES

CHAPTER 1

VIBRATIONS OF A SINGLE-DEGREE-OF-FREEDOM SYSTEM

1–1. Free Vibrations. Consider a simple spring-mass system as shown in Fig. 1–1. The spring is assumed to exhibit a force directly proportional to its displacement. Under the action of the weight W the spring will deflect some amount, called the *static displacement,* defined by

$$k\delta_{st} = W = mg \tag{1-1}$$

where k is a proportionality constant called the *spring constant* and obviously has the dimensions of force/length; m is the mass of the body, and g is the acceleration due to gravity.

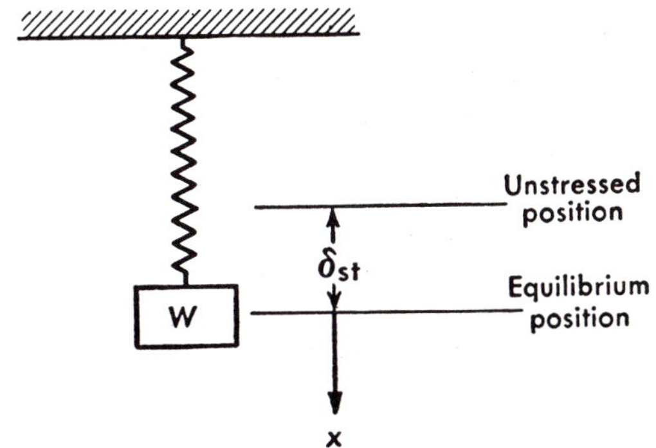

Fig. 1–1. Simple spring-mass system.

Taking the x-axis as positive down and its origin at the position of static equilibrium, we may write Newton's second law of motion in the form*

$$m\ddot{x} = -kx \tag{1-2}$$

If we introduce the notation

$$\frac{k}{m} = \omega^2 \tag{1-3}$$

* In this text, dots over a quantity always indicate differentiation with respect to time. In Eq. 1–2 the mass of the spring is assumed to be negligible compared with the mass of the weight W.

Eq. 1–2 becomes

$$\ddot{x} + \omega^2 x = 0 \tag{1–4}$$

This is the differential equation of motion for *free* vibrations of the weight inasmuch as there are no external forces acting. The equation is linear, homogeneous, and of second order; the general solution is readily found to be (as may be verified easily by direct substitution):

$$x = A \cos \omega t + B \sin \omega t \tag{1–5}$$

where A and B are arbitrary constants which may be used to adapt the solution to any given set of initial conditions.

Assume that the weight has been set into motion by giving to it some initial displacement x_0 and releasing it at time $t = 0$ without any velocity. From Eq. 1–5 and its first derivative, we find $A = x_0$, $B = 0$. The solution then becomes

$$x = x_0 \cos \omega t$$

It may be recalled from physics that this result represents a *simple harmonic motion.* The maximum amplitude of the ensuing vibration is obviously x_0; this is the *amplitude of vibration.* From a consideration of the cosine function, we see that the weight will undergo one complete cycle of vibration during a period of time equal to $2\pi/\omega$. This length of time, called the *period of vibration,* is usually denoted by τ:

$$\tau = \frac{2\pi}{\omega} = 2\pi \sqrt{\frac{m}{k}} = 2\pi \sqrt{\frac{\delta_{st}}{g}} \tag{1–6}$$

The quantity ω appearing in the above equation has the dimensions of 1/time and may therefore be interpreted as an angular velocity. Thus, if

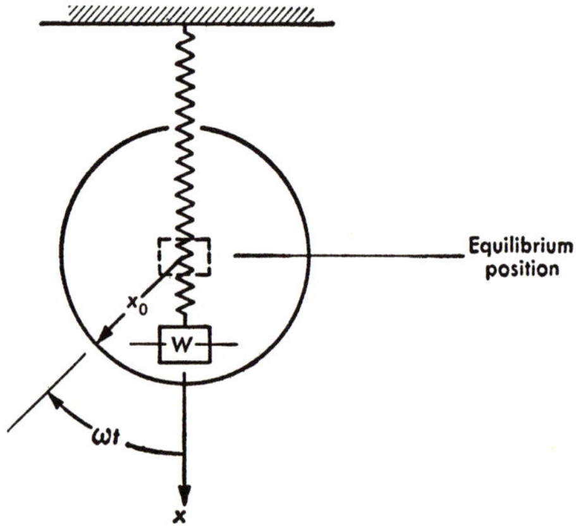

FIG. 1–2. Vector representation for simple harmonic motion.

we consider a vector of length x_0 rotating with constant angular velocity ω, the component of the vector given by $x_0 \cos \omega t$ will represent the motion of the weight along the x-axis (Fig. 1–2). In view of these remarks we see that the quantity ω is indicative of the number of cycles of oscillation performed per unit of time by the weight, and it is therefore called the *natural angular frequency* of the system, or more briefly, the *natural frequency*.

Now that we have expressed the time required for one complete cycle of vibration, we may write the number of cycles of vibration per unit of time as

$$f = \frac{1}{\tau} = \frac{\omega}{2\pi} = \frac{1}{2\pi}\sqrt{\frac{k}{m}} = \frac{1}{2\pi}\sqrt{\frac{g}{\delta_{st}}} \tag{1–7}$$

The quantity f is called the *frequency of vibration*.

Let us now examine the more general case of the weight being displaced by an arbitrary amount x_0 and released with an arbitrary initial velocity v_0 at time $t = 0$. From the previously assumed form of solution, Eq. 1–5, we find $A = x_0$, $B = v_0/\omega$. The general solution of the equation of motion thus becomes

$$x = x_0 \cos \omega t + \frac{v_0}{\omega} \sin \omega t \tag{1–8}$$

The motion described by this equation is again of simple harmonic type.

The general expression for velocity may then be obtained directly from this result by differentiation:

$$\dot{x} = v = -x_0\omega \sin \omega t + v_0 \cos \omega t \tag{1–9}$$

1–2. Free Vibrations with Damping. Imagine that the spring-mass system of the preceding article is also subjected to a force which resists the motion in a manner directly proportional to the first power of the velocity. Such a force is usually termed a *viscous* force because it represents the resistance experienced by a body moving slowly through a

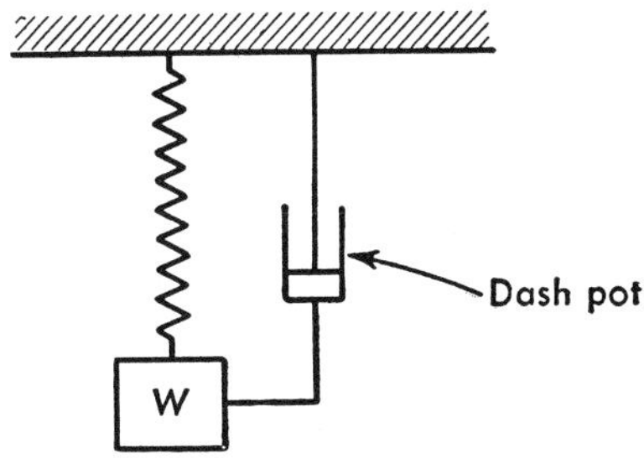

FIG. 1–3. Simple spring-mass system with viscous damping.

fluid. This is shown schematically by placing a *dash pot* between the mass and the support (Fig. 1–3). Under this additional force, the differential equation of motion becomes

$$m\ddot{x} = -kx - c\dot{x} \tag{1–10}$$

where c is a constant representing the viscous resistance at unit velocity and has the dimensions force $\times$ time/length. Defining,

$$\frac{c}{m} = 2n \tag{1–11}$$

Eq. 1–10 becomes

$$\ddot{x} + 2n\dot{x} + \omega^2 x = 0 \tag{1–12}$$

This is again a linear, homogeneous, differential equation of second order with constant coefficients. As a trial solution, assume

$$x = Ce^{pt}$$

Introducing this assumed solution into Eq. 1–12, we obtain the characteristic equation*

$$p^2 + 2np + \omega^2 = 0$$

From this equation it is seen that there are two values of p for which the trial solution will satisfy the differential equation. These are

$$p_1 = -n + \sqrt{n^2 - \omega^2}$$

$$p_2 = -n - \sqrt{n^2 - \omega^2}$$

Therefore, the general solution of the differential equation may be written as

$$x = C_1 e^{p_1 t} + C_2 e^{p_2 t} \tag{1–13}$$

where C_1 and C_2 are again two arbitrary constants which may be used to adapt the solution to any given set of initial conditions.

A closer scrutiny of the expressions for p_1 and p_2 reveals that there are apparently three subcases involved. These depend upon the relative magnitudes of n and ω. Let us try to attach physical meaning to the solution (Eq. 1–13) by examining these subcases.

1. $n > \omega$. In this instance the radicals of p_1 and p_2 are real. Furthermore, p_1 and p_2 will both be negative since $|\sqrt{n^2 - \omega^2}| < n$. Therefore, as time goes on (i.e., time becomes very large) both terms of Eq. 1–13 approach zero. The displacement-time diagram for such a motion is shown in Fig. 1–4. This motion is not really a vibration at all but rather is simply a "creeping back" of the mass from the initial disturbance.

* Also known as the *auxiliary* equation.

Such a motion is known as *aperiodic motion*, or *pure subsidence*. This type of motion results simply from the condition that the damping force is very much larger than the restoring force of the spring.

2. $n = \omega$. For this case the radicals are zero and again p_1 and p_2 are real and negative, so that this case also represents aperiodic motion. In this case the value of the damping coefficient is simply given by

$$c_c = 2nm = 2\sqrt{km} \tag{1-14}$$

This particular value of c (i.e., c_c) is called the *critical damping* coefficient because of its borderline significance between two other cases.

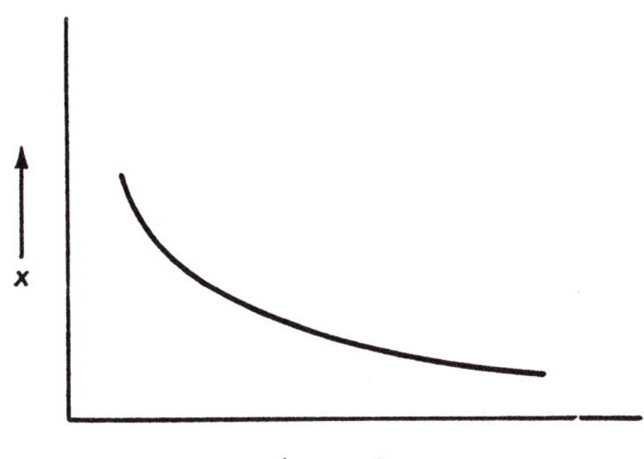

FIG. 1–4. Aperiodic motion.

3. $n < \omega$. Here we encounter the most common condition; p_1 and p_2 are complex quantities because both radicals are imaginary. Suppose that we define

$$\omega^2 - n^2 = \bar{\omega}^2 \tag{1-15}$$

Then

$$p_1 = -n + i\bar{\omega}$$

$$p_2 = -n - i\bar{\omega}$$

where $i = \sqrt{-1}$, and the general solution (Eq. 1–13) becomes

$$x = e^{-nt}(C_1 e^{i\bar{\omega}t} + C_2 e^{-i\bar{\omega}t}) \tag{1-16a}$$

Using the trigonometric identity

$$e^{\pm i\theta} = \cos\theta \pm i\sin\theta$$

we may write Eq. 1–16a in the alternate form

$$x = e^{-nt}(C_1' \cos\bar{\omega}t + C_2' \sin\bar{\omega}t) \tag{1-16b}$$

where C_1' and C_2' are new arbitrary constants. The fact that C_2' is imaginary is of no particular consequence inasmuch as C_1' and C_2' are arbitrary constants.

The terms in parentheses in Eq. 1–16b are oscillatory in character; however, they are multiplied by a factor which diminishes in value with increasing time. Therefore, the resulting motion is vibratory but with gradually diminishing amplitude. This is shown in Fig. 1–5.

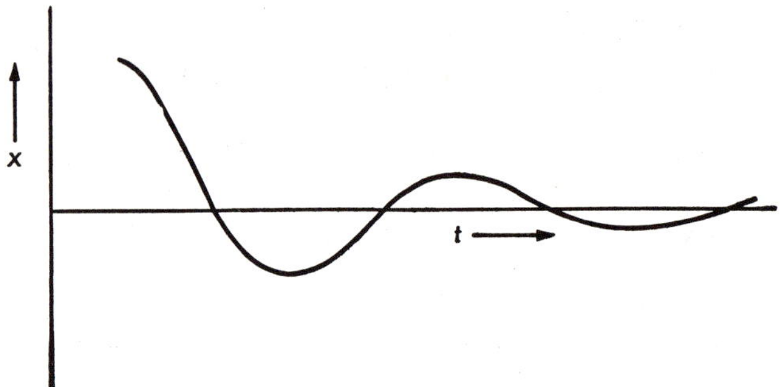

FIG. 1–5. Damped oscillatory motion.

Suppose that we were to redefine C_1' and C_2' in terms of two new constants C' and α. In particular, let

$$\begin{aligned} C_1' &= C' \sin \alpha \\ C_2' &= C' \cos \alpha \end{aligned} \tag{1–17}$$

Introducing these relations into Eq. 1–16b, we have

$$x = C'e^{-nt}(\sin \alpha \cos \bar{\omega}t + \cos \alpha \sin \bar{\omega}t)$$

or

$$x = C'e^{-nt} \sin (\bar{\omega}t + \alpha) \tag{1–16c}$$

This form of the general solution is completely equivalent to those given by Eqs. 1–16a and 1–16b.

From Eq. 1–16c we see that the loci of maximum amplitudes are given by the expression $x = \pm C'e^{-nt}$. These *envelopes* are shown in Fig. 1–6; they determine the amplitude of vibration of the damped motion. The time required to complete one cycle of vibration has the value $\tau = 2\pi/\bar{\omega}$ and is called the *period of damped vibration.* As before, the frequency of vibration is the reciprocal of the period, i.e., $f = 1/\tau = \bar{\omega}/2\pi$.

The constant α has been given the special name *phase angle* because it describes the difference in time between which x would be zero if there were no damping and the time at which x actually is zero. Or, referring to Fig. 1–6, α determines the difference in time between the first point of tangency of the displacement-time curve with the amplitude envelope and the initial time t_0. Thus the motion tends to "lag" somewhat due to the action of the viscous force.

The rate of decay of the amplitude depends upon the value of n. This rate can be calculated simply by observing the amplitudes at two points

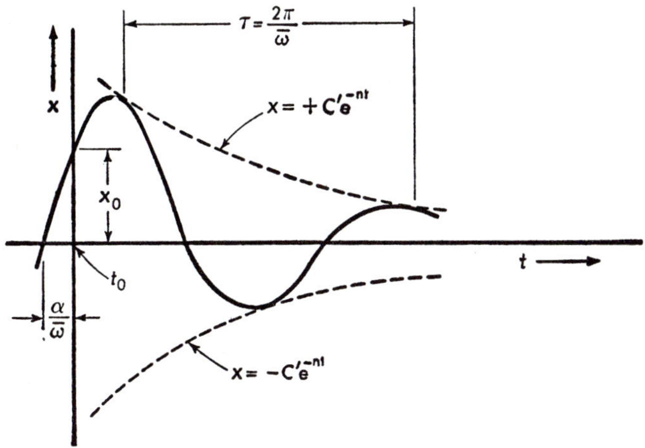

Fig. 1–6. Envelopes of damped motion.

of tangency with the envelopes. Let r denote the number of complete cycles occurring between the two points selected. Then, from Eq. 1–16c,

$$\frac{x_0}{x_r} = e^{rn\tau}$$

or

$$\log_e \frac{x_0}{x_r} = rn\tau \tag{1–18}$$

The quantity $\log_e(x_0/x_r)$ is known as the *logarithmic decrement* of the motion. The logarithmic decrement is extremely useful if one wishes to determine the damping coefficient c for a particular system from experimental observations.

Let us consider the evaluation of the arbitrary constants for the initial conditions $x = x_0$, $\dot{x} = v_0$ at $t = 0$. From Eq. 1–16b we obtain

$$C_1' = x_0, \quad C_2' = \frac{1}{\bar{\omega}}(v_0 + nx_0)$$

From these results and Eq. 1–17, we have

$$\alpha = \tan^{-1} \frac{x_0 \bar{\omega}}{v_0 + n x_0}$$

$$C' = \sqrt{x_0^2 + \frac{1}{\bar{\omega}^2} (v_0 + n x_0)^2}$$

For the initial conditions stated, the general solution of the equation of motion (Eq. 1–12) may be written in either of the two equivalent forms:

$$x = e^{-nt} \left[x_0 \cos \bar{\omega} t + \frac{1}{\bar{\omega}} (v_0 + n x_0) \sin \bar{\omega} t \right] \tag{1–19a}$$

$$x = e^{-nt} \sqrt{x_0^2 + \frac{1}{\bar{\omega}^2} (v_0 + n x_0)^2} \sin \left(\bar{\omega} t + \tan^{-1} \frac{x_0 \bar{\omega}}{v_0 + n x_0} \right) \tag{1–19b}$$

Note that the free vibrations of a simple spring-mass system with damping are completely determined by the amplitude, period, and phase angle.

1–3. Vibrations Excited by an Arbitrary External Force. Suppose that we consider the weight in Fig. 1–3 to be subjected to a small impulse at some time $t = t_i$. If the magnitude of the force (per unit mass) causing the impulse is denoted by q_i, and the duration of the impulse is taken as the small increment of time Δt_i, then the weight W will receive an instantaneous increase in velocity given by

$$\Delta \dot{x}_i = q_i \, \Delta t_i \tag{1–20}$$

This result is easily seen if we write Newton's second law in the form

$$F = \frac{d}{dt} (mv)$$

where F is the force and v is the velocity. Or,

$$F \, dt = d(mv)$$

which is the well-known *impulse-momentum theorem.** With m constant, and by taking finite (but very small) increments, this becomes

$$\Delta \dot{x} = \frac{F}{m} \Delta t$$

as above.

The general expression for the free vibrations of the system of Fig. 1–3 was given by Eq. 1–19a as

$$x = e^{-nt} \left[x_0 \cos \bar{\omega} t + \frac{1}{\bar{\omega}} (v_0 + n x_0) \sin \bar{\omega} t \right] \tag{1–21}$$

* See, *e.g.*, Dan H. Pletta, *Engineering Statics and Dynamics* (New York: The Ronald Press Co., 1951), p. 299.

Treating the incremental velocity of Eq. 1–20 as the initial velocity of the motion at the time $t_i(x_0 = 0)$, we may write the displacement at a later time t from Eq. 1–21 in the form

$$x_{t-t_i} = \Delta x = e^{-n(t-t_i)} \frac{q\,\Delta t_i}{\bar{\omega}} \sin \bar{\omega}(t - t_i) \tag{1–22}$$

Note that this result gives the incremental displacement resulting from an impulse occurring at the time t_i; the total displacement x from the time $t = 0$, then, is the sum of Eqs. 1–21 and 1–22.

Now suppose that we subject the weight to a series of successive impulses such as that just treated. If the series of k impulses is begun at time $t = 0$ and continued to time t, we may add up all the incremental displacements thereby occurring; i.e., that portion of the total displacement due to the successive impulses may be written as

$$\sum_{i=1}^{k} \Delta x_i = \frac{1}{\bar{\omega}} \sum_{i=1}^{k} q_i(t_i) e^{-n(t-t_i)} \sin \bar{\omega}(t - t_i)\,\Delta t_i \tag{1–23}$$

A series of k impulses, $q_i\,\Delta t_i$, may represent, if the duration of the impulse Δt_i is taken vanishingly small, a continuous variation of force with time. Therefore, we have succeeded in treating the case of a spring-supported mass subjected to an arbitrary external force.

The differential equation describing such a situation is of second order and linear but nonhomogeneous:

$$\ddot{x} + 2n\dot{x} + \omega^2 x = q(t) \tag{1–24}$$

The complete solution of this differential equation is, in view of the preceding analysis,

$$x = e^{-nt}\left[x_0 \cos \bar{\omega}t + \frac{1}{\bar{\omega}}(v_0 + nx_0) \sin \bar{\omega}t\right] + \frac{1}{\bar{\omega}} \int_0^t q(t_i) e^{-n(t-t_i)} \sin \bar{\omega}(t - t_i)\,dt_i \tag{1–25}$$

The integral in this equation is of extreme importance in the study of dynamics; it is known in the mathematical literature as *Duhamel's integral.* This integral, then, represents a particular solution of the differential equation (Eq. 1–24); since the first two terms of the solution represent the solution of the corresponding homogeneous equation, and hence the *free vibrations*, it is clear that the Duhamel integral represents that portion of the motion resulting from the externally applied force. Various methods, both numerical and graphical, have been devised in order to evaluate this integral; some of these methods may be found in References 4–6 at the end of this chapter. In later chapters we shall see that Duhamel's integral appears very frequently in airplane dynamic problems.

1–4. Vibrations Excited by a Harmonic Disturbing Force. In the preceding article, the vibrations resulting from an arbitrary external force, acting as a function of time, were investigated. Let us now specialize that analysis to the important case where the external force is periodic in time. The nonhomogeneous term in the differential equation (Eq. 1–24) may then be written

$$q(t) = \frac{Q_0}{m} \cos \Omega t = Q \cos \Omega t$$

where Q_0 is a constant which describes the maximum value of the forcing function, and Ω is the frequency at which the disturbing force is applied. The applicable differential equation is then

$$\ddot{x} + 2n\dot{x} + \omega^2 x = Q \cos \Omega t \qquad (1\text{–}26)$$

This is a rather simple type of nonhomogeneous differential equation, a *particular* solution of which is readily seen to be

$$x_p = A_1 \cos \Omega t + B_1 \sin \Omega t \qquad (1\text{–}27)$$

If we determine the constants A_1 and B_1 from Eq. 1–26 by direct substitution, and consider for the moment that there is no damping in the system (i e., $n = 0$), we have

$$x_p = \frac{Q}{\omega^2 - \Omega^2} \cos \Omega t$$

This result may also be obtained directly from Duhamel's integral:

$$x_p = \frac{1}{\omega} \int_0^t q(t_i) \sin \omega(t - t_i)\, dt_i$$

$$= \frac{1}{\omega} \int_0^t Q \cos \Omega t_i \sin \omega(t - t_i)\, dt_i$$

$$= \frac{Q}{\omega^2 - \Omega^2} (\cos \Omega t - \cos \omega t)^*$$

The *complete* solution (with zero damping) would then be (using Eq. 1–5)

$$x = A \cos \omega t + B \sin \omega t + \frac{Q}{\omega^2 - \Omega^2} \cos \Omega t$$

* The term cos ωt is superfluous here, since it is already contained in the solution of the corresponding homogeneous differential equation.

The complete solution of the differential equation (Eq. 1–26) *with damping* may therefore be written as the sum of the homogeneous solution (Eq. 1–19a) and the particular solution (Eq. 1–27):

$$x = e^{-nt}[A \cos \bar{\omega} t + B \sin \bar{\omega} t] + A_1 \cos \Omega t + B_1 \sin \Omega t \tag{1–28}$$

The evaluation of the constants A_1 and B_1 (note that these constants are not arbitrary and cannot be evaluated from the initial conditions but rather depend upon the differential equation itself) may be performed as follows: Introduce the solution (Eq. 1–27) into the differential equation (Eq. 1–26):

$$(-A_1\Omega^2 + 2nB_1\Omega + \omega^2 A_1 - Q) \cos \Omega t + (-B_1\Omega^2 - 2nA_1\Omega + \omega^2 B_1) \sin \Omega t = 0$$

If this equation is to be valid for all values of the time, the quantities within the parentheses must vanish simultaneously. Therefore,

$$-A_1\Omega^2 + 2nB_1\Omega + \omega^2 A_1 - Q = 0$$

$$-B_1\Omega^2 - 2nA_1\Omega + \omega^2 B_1 = 0$$

Upon solving for A_1 and B_1, we obtain

$$A_1 = \frac{Q(\omega^2 - \Omega^2)}{(\omega^2 - \Omega^2)^2 + 4n^2\Omega^2}$$

$$B_1 = \frac{2Qn\Omega}{(\omega^2 - \Omega^2)^2 + 4n^2\Omega^2} \tag{1–29}$$

Let us now study the physical interpretation for the case of zero damping. The solution of the equation of motion is

$$x = A \cos \omega t + B \sin \omega t + \frac{Q}{\omega^2 - \Omega^2} \cos \Omega t \tag{1–30}$$

The first two terms on the right-hand side represent the free vibrations while the third term, above, represents the effect of the harmonic disturbing force. The resulting motion is then a harmonic oscillation made up of the sum of two others.

It is true that Eq. 1–30 holds exactly only for the case of zero damping; however, every physical system possesses some damping, no matter how small the amount may be. In that case the terms involving ωt in Eq. 1–30 are multiplied by e^{-nt} and will therefore disappear after some time has elapsed, leaving the result

$$x = \frac{Q}{\omega^2 - \Omega^2} \cos \Omega t \tag{1–31}$$

What we have done, physically, is to allow a very small amount of damping in the system to dissipate the free vibrations so that after a long time we shall be left with only the steady-state forced vibrations. The *transient* free vibrations no longer exist.

It is readily seen from Eq. 1–31 that the period of the forced vibration is identical with the period of the disturbing force, and that the forced vibration is indefinitely sustained by the disturbing force. Furthermore, if we look for the maximum value of the displacement x, we see that it will depend upon the factor $\omega^2 - \Omega^2$. Rewrite Eq. 1–31 as

$$x = \frac{Q/\omega^2}{1 - \dfrac{\Omega^2}{\omega^2}} \cos \Omega t$$

Introducing the notation

$$\frac{\Omega^2}{\omega^2} = \eta^2 \tag{1–32}$$

and recalling that

$$Q = Q_0 \frac{g}{W} = \frac{\omega^2}{k} Q_0$$

we have

$$x = \frac{Q_0/k}{1 - \eta^2} \cos \Omega t \tag{1–33}$$

The factor $\left|\dfrac{1}{1 - \eta^2}\right|$ appearing above is called the *magnification factor*, inasmuch as it defines the amount by which the basic oscillation is increased over that which would exist if the forcing frequency were zero (i.e., if the external force were applied statically). It is therefore of interest to study the variation of the magnification factor with the frequency ratio η^2. For $\eta^2 = 0$, the value of the factor is unity; for η^2 very large, the value of the factor approaches zero. Of extreme importance is the fact that when η^2 approaches unity the factor increases without limit. The variation of magnification factor with the square of the frequency ratio is shown in Fig. 1–7.

The condition $\omega^2 = \Omega^2$ at which the magnification factor becomes infinite is called the *resonance condition.* In other words, resonance occurs when the frequency of the exciting force is equal to the natural frequency of the system. It is this phenomenon for which the study of vibration theory is so very important; the circumstance whereby a system may undergo oscillations of extremely large amplitude even when subjected to a disturbing force of very small intensity is obviously one to be

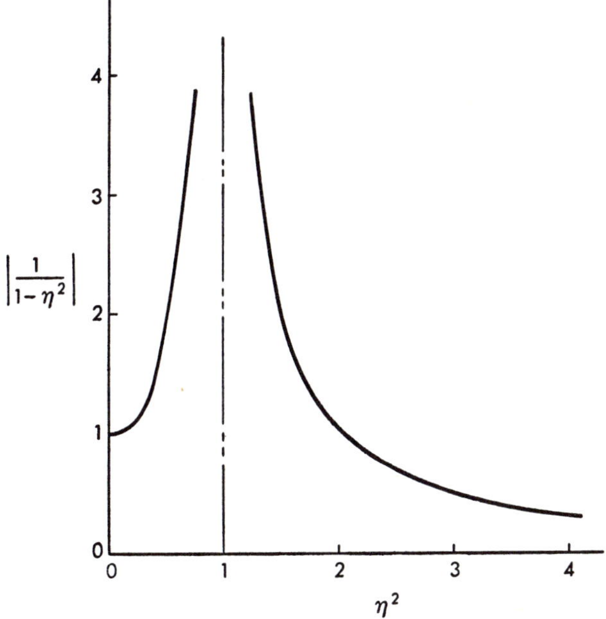

FIG. 1–7. Plot of magnification factor vs. frequency ratio for forced vibration of an undamped system.

avoided at all costs. At the resonance condition the amplitude of vibration increases uniformly with time, as shown in Fig. 1–8.

Further study of Eq. 1–33 and Fig. 1–7 will reveal the following important observations: For values of $\eta^2 < 1$, the forced oscillation is in "phase" with the disturbing force (i.e., both force and displacement

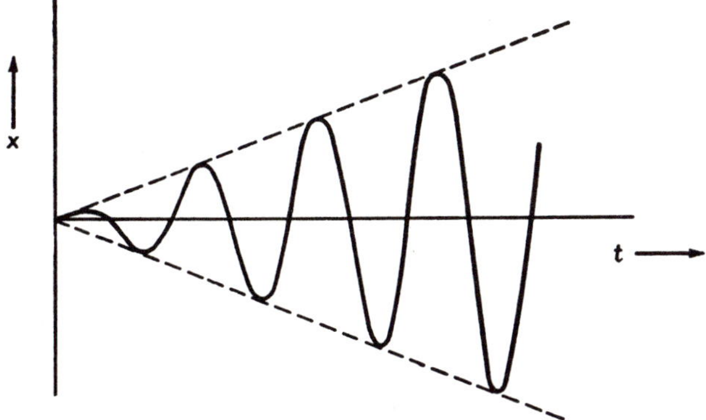

FIG. 1–8. Growth of amplitude at resonance.

have the same sign at the same time), while for values of $\eta^2 > 1$, the forced oscillation is in "opposite phase" to the disturbing force. For $\eta^2 \to 0$, the system behaves as if it were subjected to a static force Q_0.

The curve shown in Fig. 1–7 is called a *response* curve because of its representation of the amplitude of vibration of a system in response to an external stimulus.

Now let us consider the situation when damping is not neglected. The particular solution (Eq. 1–27) may be written in an alternate form, involving a phase angle α_1, as

$$x = C_1 \cos(\Omega t - \alpha_1) \tag{1–34}$$

where C_1 and α_1 are new constants given by

$$C_1 = \sqrt{A_1^2 + B_1^2} = \frac{Q_0/k}{\sqrt{(1 - \eta^2)^2 + \dfrac{4n^2}{\omega^2}\eta^2}}$$

$$\alpha_1 = \tan^{-1}\frac{B_1}{A_1} = \tan^{-1}\frac{(2n/\omega)\eta}{1 - \eta^2} \tag{1–35}$$

Again, the damping will cause the transient (free) vibrations to disappear so that we are left with the steady-state forced vibrations in the form

$$x = \frac{Q_0/k}{\sqrt{(1 - \eta^2)^2 + (4n^2/\omega^2)\eta^2}} \cos\left\{\Omega t - \tan^{-1}\left[\frac{2(n/\omega)\eta}{1 - \eta^2}\right]\right\} \tag{1–36}$$

The magnification factor is now given by

$$\text{MF} = \left[(1 - \eta^2)^2 + \frac{4n^2}{\omega^2}\eta^2\right]^{-1/2} \tag{1–37}$$

and depends upon the damping coefficient n as well as the frequency ratio η^2. We may again plot magnification factor versus frequency ratio squared, but now instead of a single curve, we shall have a family of curves with the damping factor n^2/ω^2 as a parameter. The response curves are shown in Fig. 1–9. Note that the essential effect of the damping is to limit the amplitudes near the resonance condition; for values of η^2 removed from unity, all the damped response curves approach the undamped response curve. The conclusion may be drawn, then, that the effects of damping are most important near the resonance condition.

The variation of the phase angle α_1 with frequency ratio is also of importance. When damping was neglected, it was found that the vibrations were either exactly in phase or exactly out of phase with the disturbing force; i.e., $\alpha = 0, \pi$. When damping is present, however, there is a continuous variation of α_1 with η^2. From Eq. 1–35 we see that for all values of n the value of α_1 at $\eta^2 = 1$ is $\pi/2$. Curves of α_1 versus η^2, with

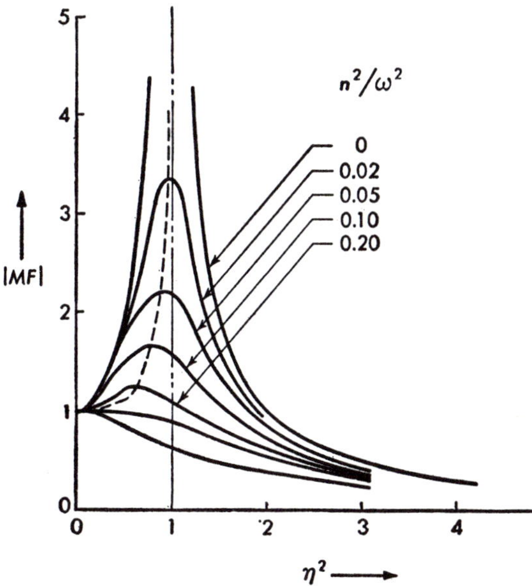

FIG. 1–9. Response curves for forced vibrations with damping.

n^2/ω^2 as a parameter (corresponding to the curves of Fig. 1–9), are given in Fig. 1–10.

It is often desirable to know the maximum displacement which the system will undergo within the entire range of forcing frequencies. From Fig. 1–9, it is seen that these maximum values, sometimes called the *resonance amplitudes*, occur very near to the resonance condition (but *at* the resonance condition, $\eta^2 = 1$, only for the case of zero damping where the value is infinite). As an approximation, however, we may take the

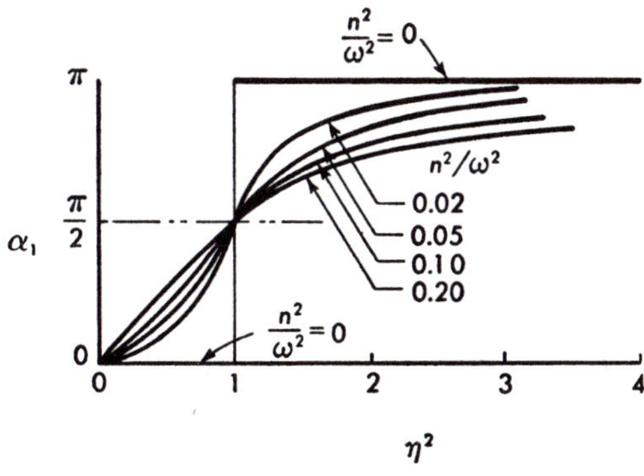

FIG. 1–10. Phase angle vs. frequency ratio for forced vibrations with damping.

resonance condition as the point at which the resonance amplitudes occur. Then, from Eqs. 1–36 and 1–37, we have the very simple relation

$$x_{\max} = \frac{Q_0}{2n\sqrt{km}} \tag{1-38}$$

A more refined analysis requires that we determine the locus of horizontal tangents of the response curves (shown dotted in Fig. 1–9). In actual practice, however, the relation (Eq. 1–38) is usually considered to give sufficient accuracy, except for very large amounts of damping.

There is yet another phenomenon occurring in the theory of vibrations which is of importance. Earlier in this article we had the expression

$$x = \frac{Q}{\omega^2 - \Omega^2}(\cos \Omega t - \cos \omega t)$$

This relation was derived under the assumption of zero damping; however, it may apply approximately to the state of combined free and forced vibrations at the beginning of motion. This relation represents the superposition of two simple harmonic vibrations of different frequency. Suppose that the two frequencies are very nearly equal; denote their difference by ϵ and their sum by σ so that

$$\omega - \Omega = \epsilon, \quad \omega + \Omega = \sigma \tag{1-39}$$

The equation above may then be written in the form

$$x = \left(\frac{2Q}{\omega^2 - \Omega^2} \sin \frac{\epsilon t}{2}\right) \sin \frac{\sigma t}{2} \tag{1-40}$$

This is a motion of period $\tau = 4\pi/\sigma$ whose amplitude varies with time and has itself a "period" $\tau' = 2\pi/\epsilon$. The latter quantity is the time during which the amplitude increases from zero to a maximum and then decreases back to zero. Thus, the motion may be pictured as shown in Fig. 1–11. This type of motion is known as *beating* because of the periodic

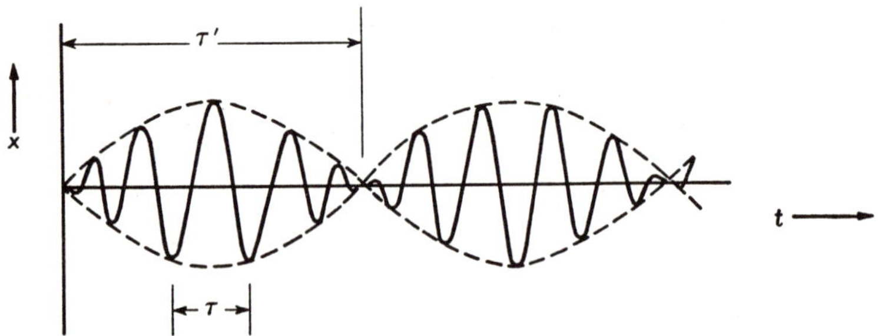

FIG. 1–11. Beating motion.

growth and decay of the vibration amplitude, as expressed by the *period of beating*, τ'. Beating is very often observed during vibration tests of airplane components.

For summing up this discussion of the forced vibration of a single-degree-of-freedom system, the following brief quotation from C. E. Inglis* will serve admirably:

> In this behavior of the spring-supported mass, there is something almost human; it objects to being rushed. If coaxed gently and not hurried too much, it responds with perfect docility; but if urged to bestir itself at more than its normal gait, it exhibits a mulish perversity of disposition. Such movement as it makes under this compulsion is always in a retrograde direction, and the more it is rushed the less it condescends to move. On the other hand, if it is stimulated with its own natural inborn frequency, it plays up with an exuberance of spirit which may be very embarrassing.

PROBLEMS

1–1. A spring-supported object weighing 7 lb is observed to vibrate with a period of $\tau = \frac{1}{2}$ sec. The damping is such that after ten complete cycles the amplitude has decreased from 3 in. to 1.5 in. Calculate the damping coefficient c.

1–2. Show that the curves of Figs. 1–9 and 1–10 may just as well have been plotted with the ratio $(c/c_c)^2$ as the parameter rather than $(n/\omega)^2$.

1–3. Consider a spring-mass system for which $c > c_c$. If motion starts at $t = 0$ with $x = 0$, $\dot{x} = v_0$, find the time at which the maximum displacement occurs and the value of the displacement at that time.

1–4. Prove that at the resonance condition in an undamped system, the displacement increases uniformly with time (see Fig. 1–8).

1–5. A weight of 5 lb is suspended from a spring having a spring contant of 1 lb/in. If the weight is raised until the tension in the spring is zero and is then released without any initial velocity, determine the period of vibration and the maximum displacement at the weight.

1–6. Find an expression for the frequency of vibration for the system shown in Fig. 1–12.

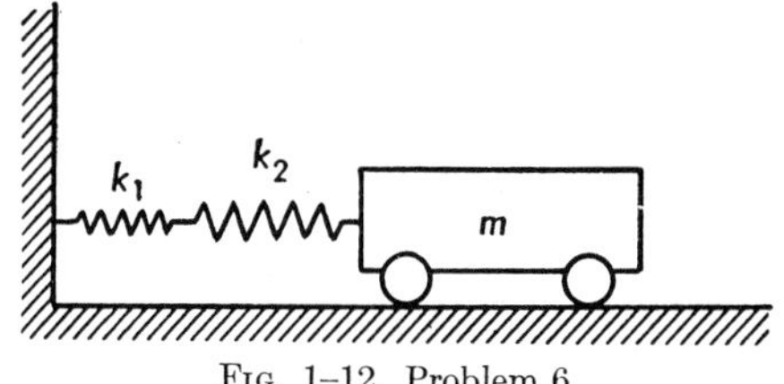

FIG. 1–12. Problem 6.

1–7. Find an expression for the frequency of vibration for the system shown in Fig. 1–13.

* C. E. Inglis, *A Mathematical Treatise on Vibrations in Railway Bridges* (New York: Cambridge University Press, 1934), p. x.

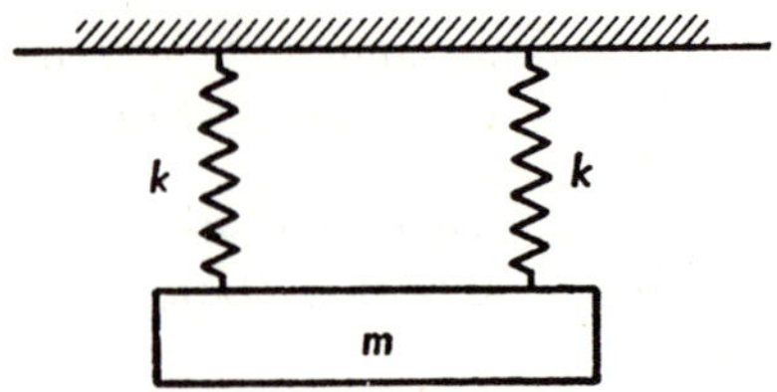

FIG. 1-13. Problem 7.

1-8. The system shown in Fig. 1-14 is observed to vibrate with a period of τ_1. When the weight W_1 is replaced by another weight W_2, the observed period is τ_2. If the weight W_1 is known, determine W_2.

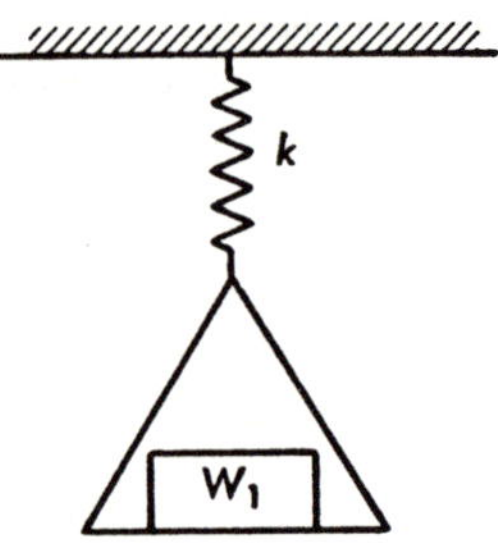

FIG. 1-14. Problem 8.

1-9. Derive the complete displacement-time equation for an undamped spring-mass system if the external disturbing force is given by

$$Q_0 = Kt$$

1-10. Derive the equation of motion for the simple pendulum shown in Fig. 1-15. Can you solve this differential equation? If not, simplify it to a form which you can solve.

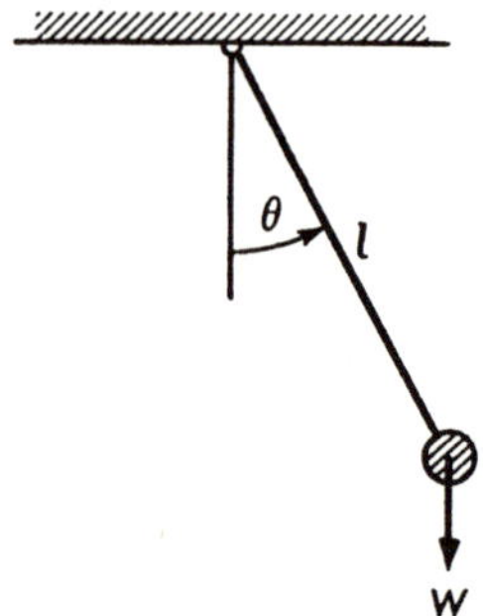

FIG. 1-15. Problem 10.

1-11. Find the frequency of vibration for each of the two systems shown in Fig. 1-16. In each case the weight of the bar is 10 lb.

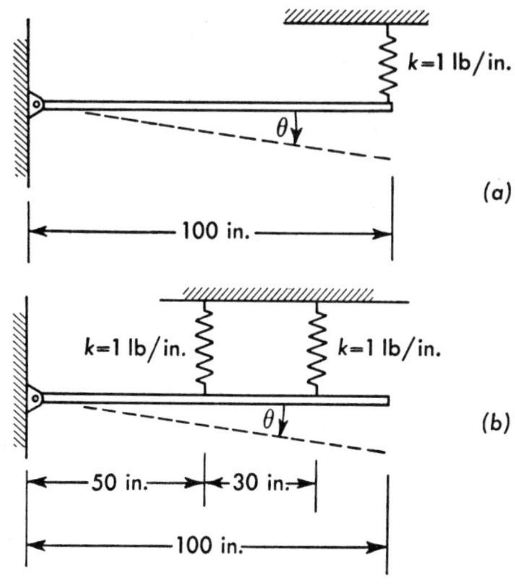

FIG. 1–16. Problem 11.

1–12. A torsional pendulum is shown in Fig. 1–17. Derive the equation of motion and find its general solution. If the initial conditions are $\theta_0 = 10°, \dot{\theta} = 0$ at $t = 0$, find θ after 2.27 sec.

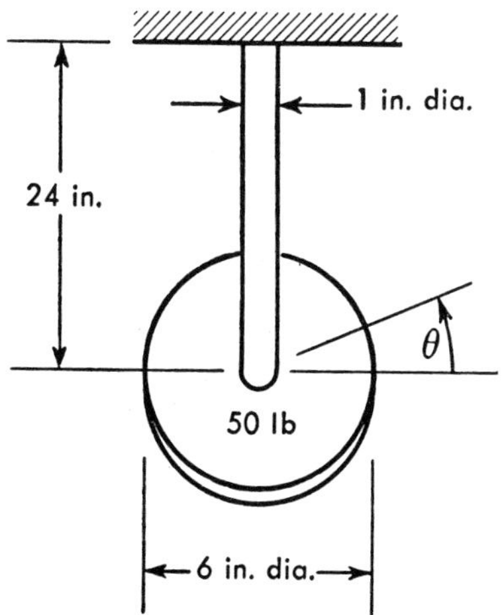

FIG. 1–17. Problem 12.

1–13. Derive an expression for the locus of the horizontal tangents of the response curves of Fig. 1–9 (shown dotted). Compare with Eq. 1–38.

1–14. For the system shown in Fig. 1–3, assume that the support undergoes a vertical displacement given by $X_0 \cos \Omega t$. Derive the equation of motion and determine the solution for steady forced vibration.

1–15. Consider the system shown in Fig. 1–18. The weight W represents a box which contains small eccentric mass m' with eccentricity e. If the mass m' rotates with an angular velocity Ω, derive the equation of motion, assuming that the system is restricted to vertical motion. Sketch the response curves.

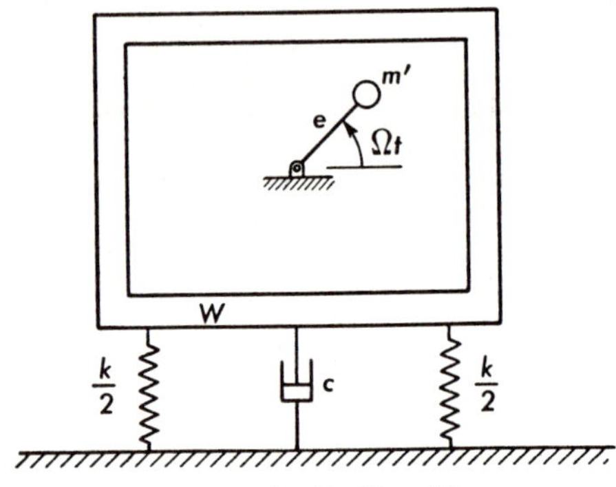

FIG. 1–18. Problem 15.

REFERENCES

1. TIMOSHENKO, S. P. *Vibration Problems in Engineering.* 3d ed.; New York: D. Van Nostrand Co., Inc., 1955.
2. DEN HARTOG, J. P. *Mechanical Vibrations.* 4th ed.; New York: McGraw-Hill Book Co., Inc., 1956.
3. THOMPSON, W. T. *Mechanical Vibrations.* 2nd ed.; Englewood Cliffs, N. J.: Prentice-Hall, Inc., 1953.
4. TIMOSHENKO, S. P., and YOUNG, D. H. *Advanced Dynamics.* New York: McGraw-Hill Book Co., Inc., 1948. Pp. 49–60.
5. VON KÁRMÁN, T., and BIOT, M. A. *Mathematical Methods in Engineering.* New York: McGraw-Hill Book Co., Inc., 1940. Pp. 403–405.
6. LAITONE, E. V., and AHLIN, J. T. "A Simple Graphical Solution of the Duhamel Integral," *Jour. Aero. Sciences,* **18** (February, 1951): 142–143.

CHAPTER 2

VIBRATIONS OF MULTI-DEGREE-OF-FREEDOM SYSTEMS

2–1. Lagrangian Equations of Motion. Consider the general motion of a particle of mass m. Let the rectangular components of the resultant force acting on the particle be X, Y, Z. Then the equations of motion resulting from the application of Newton's second law are simply

$$m\ddot{x} = X, \quad m\ddot{y} = Y, \quad m\ddot{z} = Z \tag{2–1}$$

Frequently, the systems with which we deal in mechanics involve forces the work of which are functions of the initial and final states of the system only and not of the intermediate states. Such forces are said to be derivable from a *potential function.* For example, in the simple spring-mass system of Chapter 1, the energy stored in the spring by extension is completely recoverable, the amount of energy stored depending only upon the initial and final deflections. Dynamical systems having forces derivable from such potential functions are called *conservative systems.*

It is a property of force-potential functions that their derivative in a certain direction will yield the component of the resultant force in that direction. That is, if U is the force potential representing the resultant force acting on the particle of mass m, then

$$X = -\frac{\partial U}{\partial x}, \quad Y = -\frac{\partial U}{\partial y}, \quad Z = -\frac{\partial U}{\partial z} \tag{2–2}$$

The reason for the minus sign will become apparent in the subsequent paragraphs.

The kinetic energy of the particle may be written as

$$T = \tfrac{1}{2}m(\dot{x}^2 + \dot{y}^2 + \dot{z}^2) \tag{2–3}$$

If we note that

$$\frac{\partial T}{\partial \dot{x}} = m\dot{x}, \quad \frac{\partial T}{\partial \dot{y}} = m\dot{y}, \quad \frac{\partial T}{\partial \dot{z}} = m\dot{z} \tag{2–4}$$

and write the equations of motion (Eq. 2–1) in the form

$$\frac{d}{dt}(m\dot{x}) = X = -\frac{\partial U}{\partial x}$$

$$\frac{d}{dt}(m\dot{y}) = Y = -\frac{\partial U}{\partial y}$$

$$\frac{d}{dt}(m\dot{z}) = Z = -\frac{\partial U}{\partial z}$$

we obtain

$$\frac{d}{dt}\left(\frac{\partial T}{\partial \dot{q}}\right) + \frac{\partial U}{\partial q} = 0 \tag{2–5}$$

where q is any of the coordinates x, y, z. This is a special form of *Lagrange's equation.*

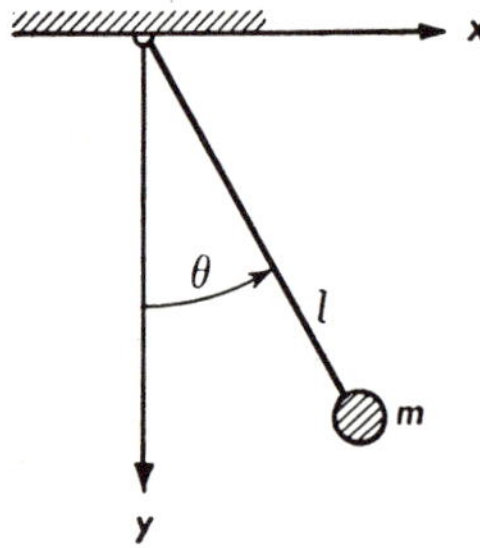

FIG. 2–1. Simple pendulum illustrating independent coordinates.

Lagrange's equation plays a very important role in dynamical theory because of its simplicity in providing the equations of motion of a system when certain energy expressions of the system are known. The arguments leading to a more rigorous derivation of Lagrange's equation are mathematically complex and somewhat subtle in nature; they are, therefore, deferred to Appendix A. We can, however, attempt at this time to discuss several important points and facts connected with Lagrange's equation.

The proper use and interpretation of Lagrange's equation requires that one be conversant with the ideas of *generalized coordinates* and *generalized forces.* Any system of completely *independent* coordinates is called generalized coordinates. For example, if the coordinates x and y are used to describe the deflected position of a simple pendulum (Fig. 2–1), it is obvious that these coordinates are not independent. They are related to each other by a physical property of the system, namely, the length of the pendulum ℓ:

$$\ell = \sqrt{x^2 + y^2}$$

If, however, the configuration is defined by the angle θ, then θ is an independent coordinate, and therefore a generalized coordinate. Quantities which limit the independence of coordinates, such as ℓ above, are called *constraints*; the preceding equation would then be an *equation of constraint.*

It is clear then that the number of generalized coordinates is equal to the number of degrees of freedom of the system. This statement characterizes what is known as a *holonomic* system. Actually, a holonomic system is one for which the equations of constraint involve coordinates but not derivatives of coordinates.

In correspondence with generalized coordinates there are generalized forces. Let $Q_r\, dq_r$ represent the work done during a small displacement of the generalized coordinate q_r. Then the quantity Q_r will be identified as the generalized force.

Our idea of force potential may now be extended somewhat. If one of the coordinates q_r is given a small increment dq_r, then the corresponding increment in potential energy must be $(\partial U/\partial q_r)\, dq_r$, and this must be equal to the negative of the work done (by the definition of potential energy):

$$\frac{\partial U}{\partial q_r}\, dq_r = -Q_r\, dq_r$$

or

$$Q_r = -\frac{\partial U}{\partial q_r} \tag{2-6}$$

The minus signs in Eq. 2–2 are thus justified.

The application of these principles to a system of particles will result in a more general form of Lagrange's equation

$$\frac{d}{dt}\left(\frac{\partial T}{\partial \dot{q}_r}\right) - \frac{\partial T}{\partial q_r} = Q_r \tag{2-7}$$

The term $\partial T/\partial q_r$ appears here in order to provide the proper terms in the equations of motion when the kinetic energy expression involves coordinates as well as time derivatives of coordinates.* If the generalized forces have potential and can therefore be found by Eq. 2–6, Lagrange's equation becomes simply

$$\frac{d}{dt}\left(\frac{\partial T}{\partial \dot{q}_r}\right) - \frac{\partial T}{\partial q_r} + \frac{\partial U}{\partial q_r} = 0 \tag{2-8}$$

This equation holds for each generalized coordinate q_r; therefore, there will be as many equations of motion as there are generalized coordinates

* See Appendix A for complete details of the derivation. (Also see References 3–7.)

(i.e., degrees of freedom). In order to obtain the equations of motion for a given system, we need have expressions only for the kinetic and potential energies of the system. The equations of motion are thereby provided; the problem of actually solving the differential equations of motion lies yet before us.

2–2. Free Vibrations of a Two-Degree-of-Freedom System. A very simple spring-mass system having two degrees of freedom is shown in Fig. 2–2. The masses of the two bodies are denoted by m_1 and m_2, and the various spring constants are k_1, k_2, and k_3 as shown. The displacements x_1 and x_2 are to be measured from the positions of static equilibrium.

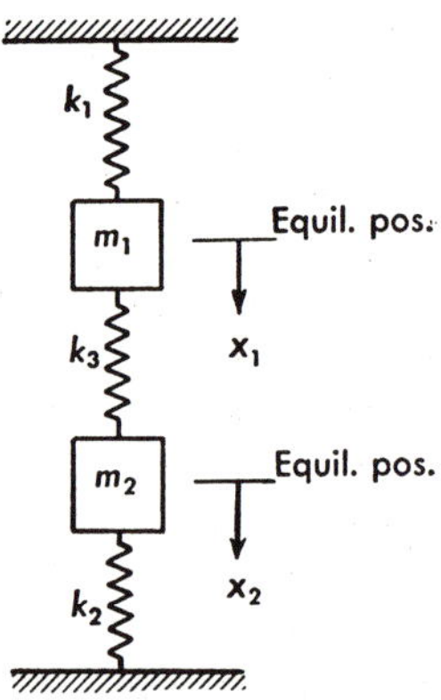

FIG. 2–2. Two-degree-of-freedom system.

The differential equations of motion for such a simple system are readily obtainable by application of Newton's second law, as was done in Chapter 1. However, in order to demonstrate the practical use of Lagrange's equation (Eq. 2–8), we shall derive the two differential equations of motion of the system by its use.

The first task is to write the kinetic and potential energies of the system. The kinetic energy is simply

$$T = \tfrac{1}{2}m_1\dot{x}_1^2 + \tfrac{1}{2}m_2\dot{x}_2^2 \tag{2–9}$$

Calculation of the potential energy requires that we have an expression for the potential energy of a spring. From our previous work (see Eqs. 1–2 and 2–6), the force exerted by the spring is*

$$Q = -\frac{\partial U}{\partial x} = -kx$$

* Note that the displacement here must be measured from the unstressed position of the spring.

from which

$$U = \frac{kx^2}{2} \tag{2-10}$$

The potential energy may therefore be written as

$$U = \tfrac{1}{2}k_1x_1^2 + \tfrac{1}{2}k_3(x_2 - x_1)^2 + \tfrac{1}{2}k_2x_2^2 \tag{2-11}$$

Substituting Eqs. 2–9 and 2–11 into Lagrange's equation (Eq. 2–8) we have the equations of motion

$$\begin{aligned} m_1\ddot{x}_1 + (k_1 + k_3)x_1 - k_3x_2 &= 0 \\ m_2\ddot{x}_2 + (k_2 + k_3)x_2 - k_3x_1 &= 0 \end{aligned} \tag{2-12}$$

These are two simultaneous, second order, homogeneous, linear differential equations. As before, solutions must be assumed for these equations, and since vibrations are involved, we try

$$\begin{aligned} x_1 &= A_1 \sin \omega t \\ x_2 &= A_2 \sin \omega t \end{aligned} \tag{2-13}$$

These represent harmonic motions of angular frequency ω and amplitudes A_1 and A_2. Introducing these assumed solutions into the equations of motion we obtain

$$\begin{aligned} A_1(-m_1\omega^2 + k_1 + k_3) - k_3A_2 &= 0 \\ -k_3A_1 + A_2(-m_2\omega^2 + k_2 + k_3) &= 0 \end{aligned}$$

Let

$$\begin{aligned} \frac{k_1 + k_3}{m_1} &= a_1, \qquad \frac{k_2 + k_3}{m_2} = a_3 \\ \frac{k_3}{m_1} &= a_2, \qquad \frac{k_3}{m_2} = a_4 \end{aligned} \tag{2-14}$$

Then we have

$$\begin{aligned} A_1(a_1 - \omega^2) - A_2a_2 &= 0 \\ -A_1a_4 + A_2(a_3 - \omega^2) &= 0 \end{aligned} \tag{2-15}$$

These are two simultaneous algebraic equations with A_1 and A_2 as unknowns.* Solutions of the type assumed (Eq. 2–13) will exist, however, only if the determinant of the coefficients

$$\begin{vmatrix} a_1 - \omega^2 & -a_2 \\ -a_4 & a_3 - \omega^2 \end{vmatrix}$$

* Actually, A_1 and A_2 are arbitrary and are to be used to satisfy the initial conditions of the motion; we should not expect to be able to solve for them in terms of the physical constants of the system.

vanishes.* Therefore, we have

$$\omega^4 - (a_1 + a_3)\omega^2 + a_1a_3 - a_2a_4 = 0 \qquad (2\text{–}16)$$

This, then, is an equation for the unknown angular frequency ω, or in other words, a *frequency equation.* We may then solve for ω^2 and obtain the two expressions

$$\begin{aligned} \omega_1^2 &= \frac{a_1 + a_3}{2} - \sqrt{\frac{(a_1 + a_3)^2}{4} - (a_1a_3 - a_2a_4)} \\ \omega_2^2 &= \frac{a_1 + a_3}{2} + \sqrt{\frac{(a_1 + a_3)^2}{4} - (a_1a_3 - a_2a_4)} \end{aligned} \qquad (2\text{–}17)$$

The radicals are always real (as may be seen by expanding through use of Eq. 2–14) and therefore the frequencies are real and positive; the system has *two* frequencies. Although the vanishing of the determinant is a necessary condition for solutions of Eq. 2–15 to exist, it is still not possible to solve for A_1 and A_2 separately; we can, however, solve for the ratio of the two; thus,

$$\frac{A_1}{A_2} = \frac{a_2}{a_1 - \omega^2} = \frac{a_3 - \omega^2}{a_4} \qquad (2\text{–}18)$$

Here there are two possible subcases depending on ω^2:

$$\begin{aligned} \left(\frac{A_1}{A_2}\right)_1 &= \frac{a_2}{a_1 - \omega_1^2} = \frac{a_3 - \omega_1^2}{a_4} \\ \left(\frac{A_1}{A_2}\right)_2 &= \frac{a_2}{a_1 - \omega_2^2} = \frac{a_3 - \omega_2^2}{a_4} \end{aligned} \qquad (2\text{–}19)$$

It is not difficult to see from Eq. 2–17 that $\omega_1^2 < a_1$ (or a_3) and $\omega_2^2 > a_1$ (or a_3) so that the ratio $(A_1/A_2)_1$ is positive and the ratio $(A_1/A_2)_2$ is negative. To demonstrate this, rewrite the radicals of Eq. 2–17 as

$$\sqrt{\frac{(a_1 - a_3)^2}{4} + a_2a_4}$$

and then write Eq. 2–17a in the form

$$a_1 - \omega_1^2 = \frac{a_1 - a_3}{2} + \sqrt{\frac{(a_1 - a_3)^2}{4} + a_2a_4}$$

Since the value of the radical is greater than $(a_1 - a_3)/2$, it is clear that $a_1 - \omega_1^2$ is positive, and therefore $\omega_1^2 < a_1$. The other cases may be proved in a similar manner.

* This is a very important theorem of algebraic equations which finds much use in the theory of vibrations. Further discussion may be found in W. L. Hart, *College Algebra* (New York: Heath, 1938), pp. 329–335.

This result shows that there are two possible *modes* of vibration of the system. In the case of vibration with the lower frequency ω_1, the amplitude ratio is positive, so that the two masses both move in the same direction at the same time. This mode of vibration is known as the *lower* or *fundamental* mode. In the case of vibration with the higher frequency ω_2, the amplitude ratio is negative, so that the masses are always moving in opposite directions (i.e., either toward each other, or away from each other). This mode of vibration is known as the higher mode. These two modes of vibration are called the *principal* modes of vibration and are characterized by the two principal frequencies ω_1 and ω_2.

From the preceding results we may now write the general solutions to the equations of motion. We may express one amplitude in terms of the other by the amplitude ratios (Eq. 2–19), and we may *consider the general motion to be made up of a combination of the principal modes.* The possibility of phase differences in the motion may be accounted for by introducing phase angles α in the solutions (Eq. 2–13). Therefore, the general solutions may be written as

$$\begin{aligned} x_1 &= (A_1)_1 \sin(\omega_1 t + \alpha_1) + (A_1)_2 \sin(\omega_2 t + \alpha_2) \\ x_2 &= \frac{a_1 - \omega_1^2}{a_2}(A_1)_1 \sin(\omega_1 t + \alpha_1) + \frac{a_1 - \omega_2^2}{a_2}(A_1)_2 \sin(\omega_2 t + \alpha_2) \end{aligned} \tag{2–20}$$

There are four arbitrary constants $(A_1)_1$, $(A_1)_2$, α_1, and α_2 which may be used to satisfy any given set of initial conditions, of which there are two (displacement and velocity) for each mass.

In the preceding problem, the coordinates x_1 and x_2 were taken as the generalized coordinates. In the more general two-degree-of-freedom system, the kinetic and potential energies (in terms of generalized coordinates q_1 and q_2) could be written as

$$\begin{aligned} T &= \tfrac{1}{2}(m_{11}\dot{q}_1^2 + 2m_{12}\dot{q}_1\dot{q}_2 + m_{22}\dot{q}_2^2) \\ U &= \tfrac{1}{2}(k_{11}q_1^2 + 2k_{12}q_1q_2 + k_{22}q_2^2) \end{aligned} \tag{2–21}$$

Using these energy expressions, one may derive the general equations of motion, and then these may be analyzed in a manner similar to that used in the preceding paragraphs.

Suppose, however, that new coordinates u_1 and u_2 are introduced according to the relations*

$$\begin{aligned} q_1 &= u_1 + u_2 \\ q_2 &= K_1u_1 + K_2u_2 \end{aligned} \tag{2–22}$$

* See S. P. Timoshenko and D. H. Young, *Advanced Dynamics* (New York: McGraw-Hill Book Co., Inc., 1948), pp. 263–265.

where K_1 and K_2 are constants so chosen that

$$q_1q_2 = 0, \quad \dot{q}_1\dot{q}_2 = 0$$

Then the energy expressions in terms of u_1 and u_2 will not contain terms involving cross-products, and the equations of motion will be of the form

$$\begin{aligned} m'_{11}\ddot{u}_1 + k'_{11}u_1 &= 0 \\ m'_{22}\ddot{u}_2 + k'_{22}u_2 &= 0 \end{aligned} \tag{2-23}$$

These two equations are completely independent of each other; therefore, the two-degree-of-freedom system has been reduced to two independent single-degree-of-freedom systems. Coordinates chosen in this manner are called *normal* coordinates, and the corresponding motions are called *normal modes* of vibrations.

The use of the normal mode concept greatly facilitates the analysis of many complex vibration problems. We shall, therefore, in later work make considerable use of normal vibration modes.

2–3. Forced Vibrations of a Two-Degree-of-Freedom System. In extension of the preceding discussion of the free vibrations of a two-degree-of-freedom system, let us now study the *forced* vibrations of such a system. Suppose that one of the masses (say, m_1 in Fig. 2–2) is acted upon by a force F_1 which is harmonic in time and of frequency Ω. The equations of motion will be

$$\begin{aligned} m_1\ddot{x}_1 + (k_1 + k_3)x_1 - k_3x_2 &= F_1 \cos \Omega t \\ m_2\ddot{x}_2 + (k_2 + k_3)x_2 - k_3x_1 &= 0 \end{aligned}$$

or, in the previous notation,

$$\begin{aligned} \ddot{x}_1 + a_1x_1 - a_2x_2 &= f_1 \cos \Omega t \\ \ddot{x}_2 + a_3x_2 - a_4x_1 &= 0 \end{aligned} \tag{2-24}$$

where

$$f_1 = \frac{F_1}{m_1}$$

As in previous work (Art. 1–4), let us look for a particular solution of these equations. Assume

$$x_1 = B_1 \cos \Omega t, \quad x_2 = B_2 \cos \Omega t \tag{2-25}$$

where the constants B_1 and B_2 are *not* arbitrary but depend upon the differential equations; i.e., upon the physical constants of the system.

Introducing the assumed solutions into Eq. 2–24, we find

$$B_1 = \frac{f_1(a_3 - \Omega^2)}{(a_1 - \Omega^2)(a_3 - \Omega^2) - a_2a_4} \tag{2–26a}$$

$$B_2 = \frac{f_1a_4}{(a_1 - \Omega^2)(a_3 - \Omega^2) - a_2a_4} \tag{2–26b}$$

Thus we have forced vibrations of the same frequency as the forcing function, multiplied by the magnification factors B_1/f_1 and B_2/f_1. As before (see Art. 1–4), the forced vibrations are conveniently studied by means of a response diagram.

The frequency equation (Eq. 2–16) may be written in the form

$$(a_1 - \omega^2)(a_3 - \omega^2) - a_2a_4 = 0$$

Comparing this expression with the denominators of Eq. 2–26, we see that B_1 and B_2 will become infinite when $\Omega^2 = \omega_1^2$ or $\Omega^2 = \omega_2^2$. This demonstrates the existence of resonance phenomena in the two-degree-of-freedom, forced-vibration problem.

For values of Ω^2 that are very large, both magnification factors will approach zero; B_1/f_1 will approach zero from the negative side and B_2/f_1 will approach zero from the positive side. For very small values of Ω^2 the magnification factor will approach positive finite values; $(B_1/B_2)_{\Omega^2 \to 0} \to a_3/a_4$.

With the foregoing observations in mind we may plot the magnification factors B_1/f_1, B_2/f_1 from Eq. 2–26 as functions of Ω^2. This is shown in Fig. 2–3. The two resonance conditions, one at $\Omega^2 = \omega_1^2$ and the other at $\Omega^2 = \omega_2^2$, are clearly shown. The phase relationships are self-evident, since absolute values were not used in plotting the functions.

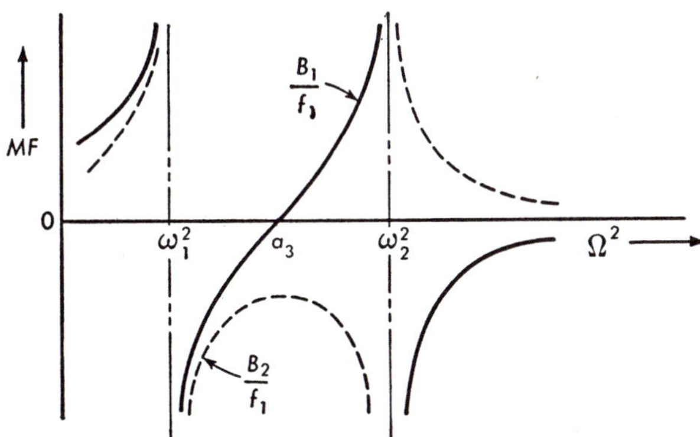

Fig. 2–3. Magnification factor vs. frequency for forced undamped vibration of a two-degree-of-freedom system.

Of special interest and importance are the conditions which exist when $\Omega^2 = a_3$. When this is so, the amplitude of vibration of m_1 will be zero (see Fig. 2–3). This phenomenon provides the essence of the *dynamic vibration absorber*. Very often one encounters the case of an elastically supported body undergoing undesirable forced vibrations. It is possible in many cases to attach a small parasitic mass to the main mass with a spring having just the right constant so that a_3 will be equal to the most commonly occurring value of Ω^2. The small parasitic mass will oscillate rather violently while the main body remains substantially at rest; in short, the parasitic mass "absorbs" the undesired vibration.

2–4. Effects of Damping. The preceding work in this chapter has omitted any considerations of the effects of damping. This article will attempt to outline in a qualitative manner the more important of these effects.

If a dash pot is placed between the two masses of Fig. 2–2, the equations of motion will become

$$\begin{aligned} m_1\ddot{x}_1 + c(\dot{x}_1 - \dot{x}_2) + (k_1 + k_3)x_1 - k_3x_2 &= F_1 \cos \Omega t \\ m_2\ddot{x}_2 + c(\dot{x}_2 - \dot{x}_1) + (k_2 + k_3)x_2 - k_3x_1 &= 0 \end{aligned} \tag{2–27}$$

where c is the dash-pot damping coefficient.

The *free* vibrations (solution of the corresponding homogeneous equations) may be found by an analysis similar to that of Art. 2–2, accounting for the damping by assuming solutions of the form

$$x_1 = A_1e^{\bar{\omega}t}, \quad x_2 = A_2e^{\bar{\omega}t} \tag{2–28}$$

where A_1 and A_2 are arbitrary constants. The results thereby obtained will correspond very closely to the results for the undamped system (see Eq. 2–20). In particular, multipliers of the form e^{-nt} will be present, and also the frequencies will be different from ω_1 and ω_2 in that they will involve the damping coefficient. The existence of the principal modes of vibration is not altered. Thus, the free vibrations are given by

$$\begin{aligned} x_1 &= e^{-n_1t}(A_1)_1 \sin(\bar{\omega}_1t + \bar{\alpha}_1) + e^{-n_2t}(A_1)_2 \sin(\bar{\omega}_2t + \bar{\alpha}_2) \\ x_2 &= e^{-n_1t}\left(\frac{a_1 - \bar{\omega}_1^2}{a_2}\right)(A_1)_1 \sin(\bar{\omega}_1t + \bar{\alpha}_1) \\ &\quad + e^{-n_2t}\left(\frac{a_1 - \bar{\omega}_2^2}{a_2}\right)(A_1)_2 \sin(\bar{\omega}_2t + \bar{\alpha}_2) \end{aligned} \tag{2–29}$$

where $\bar{\omega}_1$ and $\bar{\omega}_2$ are the damped frequencies of the system, $n_1 = c/2m_1$, $n_2 = c/2m_2$, and $\bar{\alpha}_1$, $\bar{\alpha}_2$ are phase angles.

The *forced* vibrations are sought as particular solutions of the equations of motion (Eq. 2–27). Because of the presence of damping we cannot assume such simple solutions as we did in the undamped case (Eq. 2–25). Instead, we take solutions of the form

$$\begin{aligned} x_1 &= M_1 \cos \Omega t + N_1 \sin \Omega t \\ x_2 &= M_2 \cos \Omega t + N_2 \sin \Omega t \end{aligned} \tag{2–30}$$

which will then provide for the possibility of a phase lag, as was found to appear when damping was considered in a single-degree-of-freedom system. Again, since these constants are not arbitrary but depend upon the constants in the differential equations, the constants may be evaluated, thus giving us the forced vibrations.

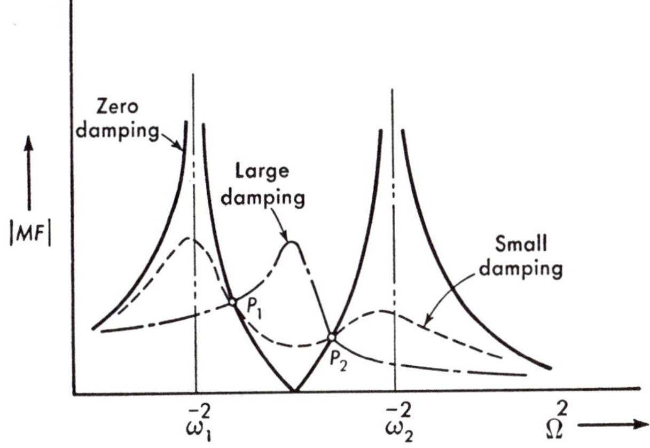

Fig. 2–4. Effect of damping on response curves for a two-degree-of-freedom system.

Actually, the analysis of the forced vibrations of the two-degree-of-freedom system with damping is quite detailed and involved. However, we may take advantage of a fact noted in connection with the study of the forced vibrations of a single-degree-of-freedom system (Art. 1–4). The fact noted there was that the principal effects of damping are to limit the vibration amplitude at the resonance condition and to shift slightly the location of the *resonance* amplitude (see Fig. 1–9). We might expect the same results to apply to multi-degree-of-freedom systems and, indeed, this is the case. Fig. 2–4 shows the response curve for the main mass m_1 for three different values of damping (zero, small, and large). Note that the absolute value of the magnification factor has been employed.

For the case of large damping it is seen that the second peak is eliminated. This results from the fact that the damping is so large that the two masses are effectively clamped together, thus resulting in essentially a single-degree-of-freedom system. In fact, for infinitely large damping between the masses, the magnification factor becomes infinite also, thereby representing an undamped single-degree-of-freedom system.

Also to be noted in Fig. 2–4 is the fact that all curves pass through the points P_1 and P_2. This is used to advantage in the *damped vibration absorber*. The object of the damped absorber is to provide a situation in which the peaks of the amplitudes occur *at* P_1 and P_2; this will then result in the minimum amplitude of vibration of the main mass over a wide range of exciting frequencies.

For a more complete discussion of the damped vibration absorber, and damped vibrations of multi-degree-of-freedom systems in general, the student is urged to consult References 1 and 2 listed at the end of this chapter.

One may wonder how equations of motion may be obtained from Lagrange's equation when damping is present. Because Lagrange's equation was derived on the basis of a conservative system, and damping forces are unconservative by their very nature (dissipation of energy), rather artificial means must be employed. Briefly, an additional term must be introduced into Lagrange's equation (Eq. 2–8) which is the derivative of a special function called a *dissipation function.*

The dissipation function for viscous damping, in terms of the generalized coordinate q_r, is simply

$$D = \tfrac{1}{2}c\dot{q}_r^2 \tag{2–31}$$

The damping force is readily obtained from the function D by differentiation; thus,

$$-\frac{\partial D}{\partial \dot{q}_r} = -c\dot{q}_r$$

Therefore, Lagrange's equation (Eq. 2–8) becomes

$$\frac{d}{dt}\left(\frac{\partial T}{\partial \dot{q}_r}\right) - \frac{\partial T}{\partial q_r} + \frac{\partial U}{\partial q_r} + \frac{\partial D}{\partial \dot{q}_r} = 0 \tag{2–32}$$

Perhaps it is worth mentioning explicitly here that if forces exist on the body which are not derivable from a potential function, then these forces appear on the right-hand side of Lagrange's equation; thus,

$$\frac{d}{dt}\left(\frac{\partial T}{\partial \dot{q}_r}\right) - \frac{\partial T}{\partial q_r} + \frac{\partial D}{\partial \dot{q}_r} + \frac{\partial U}{\partial q_r} = \bar{Q}_r \tag{2–33}$$

$\bar{Q}_r$ is the generalized force corresponding to that part of the resultant external force which is not derivable from a potential function. A *particular* solution of Eq. 2–33 would then be given by Duhamel's integral (see Art. 1–3).

2–5. Vibrations of Systems with a Finite Number of Degrees of Freedom. The analysis of the preceding articles for the two-degree-of-freedom system amply demonstrates the increased complexity resulting from the inclusion of additional degrees of freedom. It happens, moreover, that most of the problems met with in practice, especially when dealing with aircraft structures, involve more than two degrees of freedom, and therefore the analyst is confronted with problems of rather considerable complexity.

The procedure for studying a system of n degrees of freedom is, however, fairly straightforward, inasmuch as it follows the pattern used for the two-degree-of-freedom system in Art. 2–2. One must first write down the n differential equations of motion (by use of Lagrange's equation or by considering the forces acting on each mass of the system) which will have the form

$$\begin{aligned} a_1\ddot{x}_1 + b_1\ddot{x}_2 + \cdots + c_1x_1 + d_1x_2 + \cdots &= 0 \\ &\cdot \\ &\cdot \\ &\cdot \\ a_n\ddot{x}_1 + b_n\ddot{x}_2 + \cdots + c_nx_1 + d_nx_2 + \cdots &= 0 \end{aligned}$$

These equations are quite general in that they allow "coupling" to exist between the various masses, either through the inertial forces or through the elastic forces, or through both.* The problems at the end of this chapter deal with both types of coupling; furthermore, the problem of wing flutter, which is analyzed in Chapter 5, will be seen to possess coupling in the inertial and damping terms as well as the elastic terms.

As before, solutions of the following form are assumed:

$$\begin{aligned} x_1 &= A_1 \cos \omega t \\ &\cdot \\ &\cdot \\ x_n &= A_n \cos \omega t \end{aligned}$$

Introduction of the assumed solutions will then yield n algebraic equations for the unknown amplitudes $A_1 \cdots A_n$. However, in order for solutions of these algebraic equations to exist, the determinant of the coefficients must vanish.† For the two-degree-of-freedom system, the vanishing of the determinant led to a quadratic equation in ω^2 called the *frequency* equation. For a three-degree-of-freedom system, the frequency equation will be a cubic in ω^2, and for a system of n degrees of freedom, the equation will be of nth degree in ω^2.

For each root of the frequency equation we are able only to form the amplitude ratios $A_1/A_2 \cdots A_1/A_n$; but these enable us to define the

* Sometimes called *dynamic coupling* and *static coupling*, respectively.

† See footnote on page 28.

various modes of vibration corresponding to each root. Principal modes of vibration are found analogously to the two-degree-of-freedom system (Art. 2–2), and the general solution is obtained by superposition of the principal modes. There appear $2n$ arbitrary constants which may be evaluated from the $2n$ initial conditions (displacement and velocity of each element of the system).

It is actually the determination of the coefficients of the frequency equation and the solution for the roots which cause the analysis of multi-degree-of-freedom systems to be so complex. It would seem appropriate, therefore, to have available methods which would enable us to obtain the frequencies without resorting to the derivation and solution of a high-degree frequency equation. Such methods do exist, and although they give only approximate values for the frequencies, their use is widespread and exceedingly convenient. A brief discussion of some of the more important of these methods will be given in Art. 2–6.

It is now understood that the general motion of a system of n degrees of freedom is described by a superposition of the principal modes of vibration. For an exact solution we would require that *all* the modes be known. However, a general vibration composed of many principal modes is described very well by only the few lowest modes, particularly the fundamental, with modes of higher frequency altering the motion in such a manner that the higher the mode, the less its influence.* In general, what we are doing is expressing the deflections by an infinite series of terms, each term representing a mode; if the series is a convergent one, the preceding statement is an obvious consequence. Furthermore, it can be shown that the stress produced by a given load varies as the reciprocal of the frequency squared; thus, the stresses produced by the higher modes are less important than those produced by the lower modes. The proof of such statements requires the use of the concept of *orthogonal* modes of vibration† to show that the magnitude of the contribution of the various modes generally decreases with increasing order of the mode. We shall have occasion to make use of some of the properties of orthogonal modes of vibration in Chapter 7.

Therefore, in practical problems we may often be content with having information concerning only the few lowest modes. The approximate methods of determining principal frequencies to be described in the next article will therefore be mostly concerned with the determination of the lower modes of vibration. Further remarks on the importance of the higher modes and methods of determining their frequencies will be made in subsequent text discussion as we progress in the study of complex vibrating systems.

* There are exceptions, of course, but none will be discussed here.

† See Appendix B for a brief discussion of orthogonal modes.

2–6. Approximate Methods for Calculating Principal Frequencies. The general vibration problem requires that we solve an nth degree algebraic equation for the principal frequencies. This is usually difficult to do when $n > 3$; therefore, approximate methods, which obviate the necessity for solving such equations, are very often employed.

The fundamental frequency of a vibrating system is often easily calculated by consideration of the energy of the system. In a vibrating system with no damping, it is clear that the total amount of energy present is a constant. In general, part of this energy is kinetic energy due to the velocity, and part is potential energy that is stored in the elastic members. At the extreme of the motion the energy is all potential, and at the equilibrium position the energy is all kinetic; therefore, the maximum kinetic and potential energies must be equal if the total energy is to remain constant. This is the essential idea of an approximate method proposed by Lord Rayleigh.*

As an example to demonstrate the use of Rayleigh's principle, let us calculate the frequency of a simple spring-mass system if the mass of the spring is not negligible. If the weight per unit length of the spring is w and ξ is the distance to any given point in the spring from the fixed support, the displacement of that point in the spring (measured from the static equilibrium position) will be $x\xi/l$, where l is the length of the spring.† Therefore, the kinetic energy of an element of the spring will be

$$\Delta T = \tfrac{1}{2}\frac{w\Delta\xi}{g}\left(\frac{\dot{x}\xi}{l}\right)^2$$

The total kinetic energy may then be written as

$$T = \tfrac{1}{2}\frac{W}{g}\dot{x}^2 + \int_0^l \tfrac{1}{2}\frac{w}{g}\left(\frac{\dot{x}\xi}{l}\right)^2 d\xi = \frac{W}{2g}\dot{x}^2 + \frac{wl}{6g}\dot{x}^2$$

where W is the weight of the suspended mass. The potential energy stored in the spring is simply

$$U = \tfrac{1}{2}kx^2$$

Now, in order to obtain the maximum of these quantities we must have an expression for x; hence we assume

$$x = A \sin \omega t$$

Then

$$U_{\max} = \tfrac{1}{2}kA^2$$

$$T_{\max} = \left(\frac{W}{2g} + \frac{wl}{6g}\right)A^2\omega^2$$

* Rayleigh, John William Strutt, 3rd baron, *Theory of Sound*, 2nd ed. (New York: Dover Publications, 1945), pp. 111 and 287.

† This statement is strictly correct only when the weight of the spring is appreciably less than the weight of the suspended mass.

and since these two quantities are equal,

$$\omega^2 = \frac{k}{(W/g) + (wl/3g)}$$

Thus we see that in order to account for the weight of the spring, it is only necessary to add one-third of the weight of the spring to the weight of the suspended mass.

Rayleigh's method is most useful when applied to *continuous* systems (Art. 2–7).* In order to calculate the kinetic and potential energies for continuous systems, however, one must assume a *deflection shape* for the first mode; if the exact deflection shape should happen to be chosen, then the resulting frequency will be exact. It is usually sufficient to assume that the deflection shape is the same as the statical deflection shape of the system of particles acted upon by the applied loads. Rayleigh's method always gives for the fundamental frequency a value which is *too large*. This results from the fact that we place additional constraints on the system by forcing it to conform to the assumed deflection shape.†

A slight modification of the Rayleigh method consists of assuming a deflection shape which is not completely determined; i.e., some parameter is retained as an arbitrary value. Since we look for the lowest (fundamental) frequency, the value of the parameter which gives that minimum is the desired value.

A further extension of the Rayleigh method consists in specifying a deflection shape in terms of several arbitrary parameters. Again the minimum frequency is sought. This is known as the *Rayleigh-Ritz method*, or the *Ritz method*.

In the *Stodola method*, a deflection curve is assumed, and this is then changed to an inertia loading by multiplying by $m\omega^2$. Since ω^2 is not known, it is first assumed to be unity. This inertia loading is then used to obtain a new deflection curve, and so on. The frequency is found as the ratio of two successive deflection shapes. This procedure may be repeated as many times as desired, each new deflection shape giving a closer approximation to the actual frequency. Therefore, this is a method of successive approximations, or *iteration* method.

The *Holzer method* is most useful for torsional vibrations. In this method the frequency is assumed, and a deflection shape is calculated. Adjustment of the calculated deflection shape to give nodes at the proper points for the fundamental mode will then yield the fundamental frequency. The Holzer method may be considered to be the inverse of the

* The systems which we have considered so far were composed of discrete mass particles and are sometimes called *lumped-mass systems*. Systems in which the mass is distributed throughout are called *continuous systems*.

† For a mathematical proof of this statement see Reference 2, pp. 200–201, listed at the end of this chapter.

Stodola method. The Holzer method is quite useful because it may be used to determine the higher modes and frequencies.

A method similar to the Holzer method, but applied to bending vibrations, is known as the *Myklestad method* (see Reference 10). Higher modes and frequencies are fairly easily obtained by this method.

The *Matrix method* employs matrix algebra (see Chapter 3) to set up the problem in a convenient mathematical form. Influence coefficients* are used to express a *calculated* mode shape. A special procedure, known as *matrix iteration*, is used to calculate the mode frequencies. A numerical example will be given in Chapter 3 to demonstrate the essentials of the matrix iteration method.

All these methods have been very highly developed as regards computational procedures. Tabular forms have been set up which are very convenient to use and yet are flexible enough to have wide applicability. Modifications to all these methods have been proposed to provide greater accuracy and in many cases the higher modes and frequencies as well (see, for example, Reference 9). Complete discussion of these methods, and some modifications to them, can be found in the references at the end of this chapter.

2–7. Vibrations of Beams—Systems with an Infinite Number of Degrees of Freedom. The study of the vibrations of a beam forms an interesting and important part of the subject of dynamics. The vibrating system, in this case, has mass distributed throughout and is therefore classed as a continuous system. Actually, a beam is composed of a large number of individual particles (molecules), each possessing three degrees of freedom, but there are so many of these particles that we refer to the system as a continuous one possessing an infinite number of degrees of freedom.

Aircraft structures, in idealized form, may usually be resolved into components which are largely made up of beams of various types; hence the importance of beam theory in aeronautic problems. The uninitiated, looking at an airplane wing from the dynamical viewpoint for the first time, may think it more logical to treat the wing as a plate instead of as a beam. For the somewhat unconventional wing planforms being utilized for high-speed flight at the present time, this thought is perhaps very nearly correct. But for the more conventional type of wing, experience has shown that under the loadings normally encountered in flight, the wing deforms very nearly according to the beam relations. The use of beam theory for most airplane dynamics problems involving wings is therefore justified. It might be mentioned that the analysis of vibrating

* An influence coefficient is simply an expression by which one can obtain the deflection at a certain point in a structure due to the application of a unit load at some other point in the structure.

plates is not nearly so well developed as that for vibrating beams, and therefore even wings of small thickness and low aspect ratio are often treated by the conventional vibrating beam theory.

As a first example, let us determine the fundamental frequency of a weightless cantilever beam with a concentrated mass at the free end (Fig. 2–5). Rayleigh's method will be very convenient here.

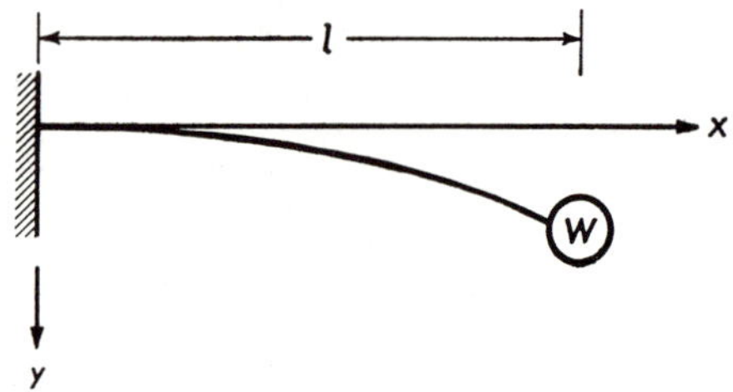

Fig. 2–5. Cantilever beam with end weight.

The static deflection of such a beam at any point x is given by*

$$y = \frac{Wx^2}{6EI}(3l - x) \tag{2–34}$$

The maximum potential energy of a beam in bending is†

$$U_{\text{max}} = \frac{EI}{2}\int_0^l \left(\frac{d^2y}{dx^2}\right)^2 dx \tag{2–35}$$

which becomes, for this problem,

$$U_{\text{max}} = \frac{W^2l^3}{6EI} \tag{2–36}$$

The kinetic energy is given by

$$T = \frac{W}{2g}\dot{y}^2 \tag{2–37}$$

Because vibrations are involved, we assume that $y(t)$ is given by

$$y = y_0 \sin \omega t \tag{2–38}$$

where ω is the frequency of vibration and y_0 the amplitude. The kinetic energy then becomes

$$T = \frac{W}{2g}\omega^2 y_0^2 \cos^2 \omega t$$

* See, e.g., S. P. Timoshenko, *Strength of Materials*, Part I (New York: D. Van Nostrand Co., 1940), p. 148.

† *Ibid.*, p. 297.

Letting y_0 be the static displacement at $x = l$, since all the kinetic energy is concentrated at this point, and taking the maximum value of the kinetic energy, we obtain

$$T_{\max} = \frac{W^3 l^6 \omega^2}{18gE^2I^2} \tag{2-39}$$

Equating maximum kinetic and potential energies (Eqs. 2–36 and 2–39) gives

$$\omega = \sqrt{\frac{3EI}{(l^3W)/g}} \tag{2-40}$$

as the frequency of the fundamental mode of vibration.

The effect of the weight of the uniform beam may easily be accounted for by introducing into the potential and kinetic energies a term involving the weight per unit length (w). The kinetic energy will now vary along the span of the beam, and therefore we require y_0 as a function of x. To find y_0, we take the static deflection curve to represent the mode shape, which then serves as an expression for the variation of amplitude y_0 along the span of the beam. The expression for the fundamental frequency will be found to be (Problem 10)

$$\omega = \sqrt{\frac{3EI}{[l^3(W + 0.236wl)]/g}} \tag{2-41}$$

We may now write a general expression for the fundamental frequency of a vibrating beam, based on Rayleigh's method. Since the maximum potential energy is

$$U_{\max} = \int_0^l \frac{EI}{2}\left(\frac{d^2y}{dx^2}\right)^2 dx \tag{2-42}$$

and the maximum kinetic energy can be written as

$$T_{\max} = \frac{\omega^2}{2g}\int_0^l wy^2\, dx \tag{2-43}$$

we obtain

$$\omega^2 = \frac{\displaystyle\int_0^l EI\,(d^2y/dx^2)^2\, dx}{\displaystyle\int_0^l \frac{w}{g} y^2\, dx} \tag{2-44}$$

Note that this expression is valid for nonuniform beams as well as for uniform ones, in which case E, I, and w will be known functions of x; therefore, the integrations can be performed, although a numerical stepwise procedure may sometimes be required.

A more elaborate analysis of the vibrating beam problem is accomplished by use of a trigonometric series to represent the deflected shape (see Reference 1). Proper selection of the generalized coordinates will then lead to a differential equation from which the principal modes may be found. For example, a simply supported uniform beam of length l

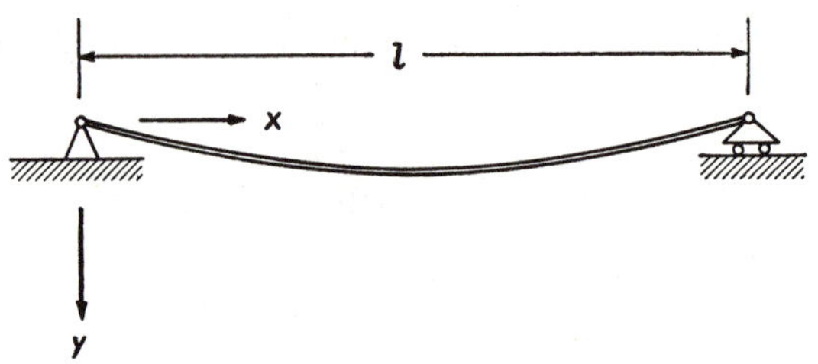

FIG. 2–6. Simply supported beam.

(Fig. 2–6) has a deflection shape which may be represented as a series of sine terms:

$$\begin{aligned} y &= \mathcal{A}_1 \sin \frac{\pi x}{l} + \mathcal{A}_2 \sin \frac{2\pi x}{l} + \cdots + \mathcal{A}_n \sin \frac{n\pi x}{l} + \cdots \\ &= \sum_{n=1}^{\infty} \mathcal{A}_n \sin \frac{n\pi x}{l} \end{aligned} \tag{2–45}$$

That this representation is a possible one may be seen easily, since $y = 0$ when $x = 0, l$ (condition of zero deflection at the supports) and $d^2y/dx^2 = 0$ when $x = 0, l$ (condition of zero moment at the supports). These conditions are called *boundary conditions.*

The quantities $\mathcal{A}_n$ in Eq. 2–45 define, at any instant, the shape of the deflection curve by their magnitude and are therefore functions of time. Let us take these quantities as the generalized coordinates of the problem. The potential energy may be written, as before, in the form

$$U_{\max} = \frac{EI}{2} \int_0^l \left(\frac{d^2y}{dx^2}\right)^2 dx \tag{2–35}$$

Differentiating Eq. 2–45 twice and squaring will result in terms of the general form

$$\frac{n^2m^2\pi^4}{l^4} \mathcal{A}_n \mathcal{A}_m \sin \frac{n\pi x}{l} \sin \frac{m\pi x}{l}, \quad \frac{n^4\pi^4}{l^4} \mathcal{A}_n^2 \sin^2 \frac{n\pi x}{l}$$

Those of the first form all integrate to zero, and those of the second form integrate to $(n^4\pi^4/2l^3)\mathcal{A}_n^2$, so that we have

$$U = \frac{EI\pi^4}{4l^3} \sum_{n=1}^{\infty} n^4 \mathcal{A}_n^2 \tag{2–46}$$

The kinetic energy is

$$T = \frac{w}{2g}\int_0^l \dot{y}^2\,dx \tag{2-47}$$

where w is the weight per unit length. Introducing Eq. 2–45, and performing the differentiation and integration, we find

$$T = \frac{wl}{4g}\sum_{n=1}^{\infty} \dot{\mathcal{A}}_n^{\,2} \tag{2-48}$$

Introducing the relations for T and U into Lagrange's equation (Eq. 2–8) yields

$$\frac{l}{2}\frac{w}{g}\ddot{\mathcal{A}}_n + \frac{EI\pi^4 n^4}{2l^3}\mathcal{A}_n = 0 \tag{2-49}$$

This is a very simple differential equation for the quantities $\mathcal{A}_n$. In fact, for each $\mathcal{A}_n$, the differential equation is identical with that for a single-degree-of-freedom system. It was noted in Art. 2–2 that such a result is obtained when the *generalized* coordinates are chosen as *normal* coordinates. Thus, each mode of vibration of the beam may be treated as an *independent* single-degree-of-freedom system.

Let

$$\frac{EI}{w/g} = \beta^4 \tag{2-50}$$

The general solution of Eq. 2–49 is then

$$\mathcal{A}_n = A_n \cos\frac{n^2\pi^2}{l^2}\beta^2 t + B_n \sin\frac{n^2\pi^2}{l^2}\beta^2 t \tag{2-51}$$

where A_n and B_n are *arbitrary* constants which may be used to satisfy the given initial conditions of the motion. The normal modes of vibration are thus described by

$$y_n = \left(A_n \cos\frac{n^2\pi^2}{l^2}\beta^2 t + B_n \sin\frac{n^2\pi^2}{l^2}\beta^2 t\right)\sin\frac{n\pi x}{l} \tag{2-52}$$

The frequency of each mode is simply

$$\omega_n = \left(\frac{n\pi}{l}\right)^2\sqrt{\frac{EI}{w/g}} \tag{2-53}$$

The general vibration is composed of the sum of the normal modes of vibration:

$$y = \sum_{n=1}^{\infty}(A_n \cos\omega_n t + B_n \sin\omega_n t)\sin\frac{n\pi x}{l} \tag{2-54}$$

The application of an external force does not introduce any particular difficulty into the analysis. All that is necessary is to determine the generalized force Q_n corresponding to each generalized coordinate $\mathfrak{a}_n$ and write Lagrange's equation according to Eq. 2–33. The *particular* solutions of the resulting differential equations may be obtained by using Duhamel's integral.

Beams with boundary conditions different from the example just treated are analyzed in a similar fashion. One must only attempt to select a representation of the deflection shape so that the boundary conditions of the particular problem are satisfied.

Another approach to the vibrating beam problem is through the use of the fundamental beam relations*

$$M = EI \frac{d^2y}{dx^2}$$
$$p = \frac{d^2M}{dx^2} \tag{2–55}$$

where M is the bending moment and p is the distributed loading. Combining these two equations gives

$$p = \frac{d^2}{dx^2}\left(EI \frac{d^2y}{dx^2}\right) \tag{2–56}$$

If the static loading p is now replaced by an inertia loading $-(w/g)(\partial^2 y/\partial t^2)$, Eq. 2–56 becomes (for a uniform beam)

$$EI \frac{\partial^4 y}{\partial x^4} = -\frac{w}{g}\frac{\partial^2 y}{\partial t^2} \tag{2–57}$$

where partial differentiation symbols have been used because y is now a function of both x and t. This fourth-order partial differential equation may be reduced to an ordinary differential equation by assuming that the deflection varies harmonically with time, i.e.,

$$y(x, t) = y(x) \sin \omega t \tag{2–58}$$

Eq. 2–57 then becomes

$$\frac{d^4y}{dx^4} - k^4 y = 0 \tag{2–59}$$

where

$$k^4 = \frac{\omega^2}{EI}\frac{w}{g} \tag{2–60}$$

* *Ibid.*, pp. 135–137.

We now have a fourth-order ordinary differential equation; the general solution of this equation is

$$y = A \sin kx + B \cos kx + C \sinh kx + D \cosh kx \qquad (2\text{–}61)$$

A, B, C, and D are arbitrary constants which may be used to account for the given boundary conditions, of which there are four. The four boundary conditions can be chosen from the deflections, slopes, shears, or moments at each end of the beam (or any other particular point along the beam which is convenient); which of these eight quantities are used for the four boundary conditions depends on the particular problem at hand. For the simply supported beam, these quantities were zero deflection and moment at each support; for a cantilever beam, these would be zero deflection and slope at the built-in end, and zero shear and bending moment at the free end.

Introduction of the boundary conditions will result in a frequency equation. For example, the cantilever-beam boundary conditions will lead to $A = -C$, $B = -D$, and the two algebraic equations

$$C\,(\sin kl + \sinh kl) + D\,(\cos kl + \cosh kl) = 0$$

$$C\,(\cos kl + \cosh kl) - D\,(\sin kl - \sinh kl) = 0$$

For solutions (other than $C = D = 0$) to exist, the determinant of the coefficients must vanish,* which then leads to the equation

$$1 + \cos kl \cosh kl = 0$$

We then seek values of k for which this equation is satisfied, and thus we find the *principal* frequencies ω_n.

All the discussion of this section has been confined to beams of constant cross-sectional properties along the length; that is, *uniform* beams. However, most beams encountered in practice have properties which vary along their length; that is, they are *nonuniform*. This variation is usually accounted for by performing all required integrations in a step-by-step manner. There are in existence many tabular methods for dealing with nonuniform beams (see References 1, 2, 8–11), some of which are widely used. Most of the approximate methods for determining frequencies and modes described in Art. 2–6 have been adapted to tabular forms and may be used for nonuniform beams. The matrix method, which uses influence coefficients, is particularly well adapted for this type of work.

PROBLEMS

2–1. Derive the complete solutions for the free vibrations of a two-mass system with damping between the masses (Eq. 2–29).

* See footnote on page 28.

2–2. Find the frequency of torsional vibration for the shaft-disk system shown in Fig. 2–7.

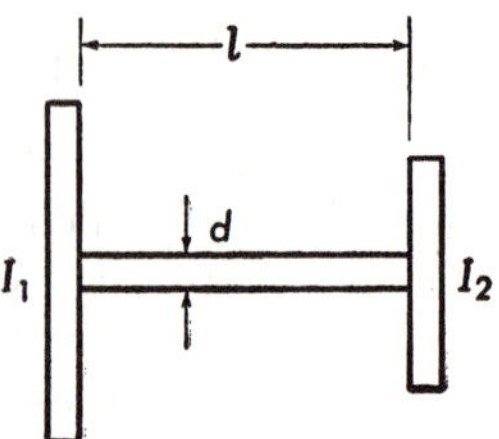

Fig. 2–7. Problems 2 and 3.

2–3. Derive the equations of motion of the disks for the system shown in Fig. 2–7.

2–4. Two identical simple pendulums are connected by a spring as shown in Fig. 2–8. Assuming the entire mass of the system to be concentrated in the two bobs, derive the two equations of motion, and find the two principal frequencies of vibration.

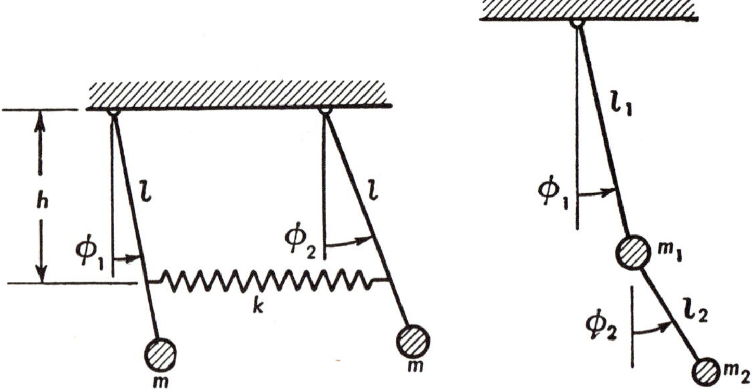

Fig. 2–8. Problem 4. Fig. 2–9. Problem 5.

2–5. Calculate the principal frequencies of vibration for the system shown in Fig. 2–9. Derive the equations of motion by use of Lagrange's equation.

2–6. For the torsional system shown in Fig. 2–10, derive the equations of motion, and find the two principal frequencies.

2–7. An airplane has been idealized to the dynamic system shown in Fig. 2–11, where

m_1 = mass of airplane
m_2 = mass of landing gear
k_1 = spring constant of shock strut
k_2 = spring constant of tire
c = damping coefficient of shock strut

Derive the equations of motion.

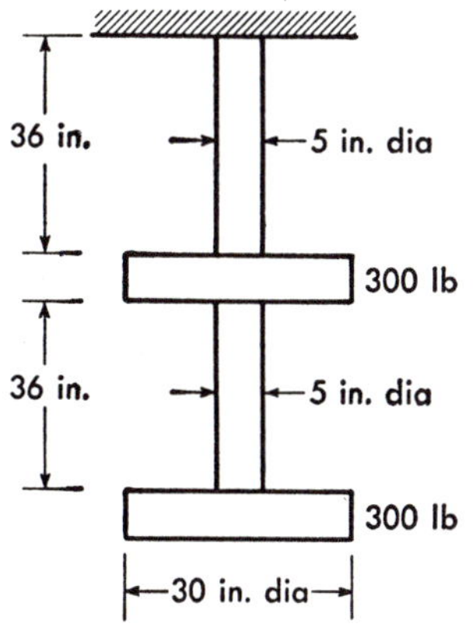

FIG. 2-10. Problem 6.

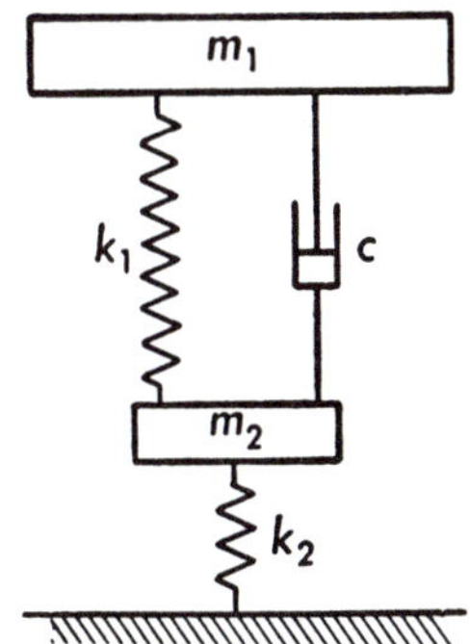

FIG. 2-11. Problem 7.

2-8. An airfoil is placed in a wind tunnel for the purpose of performing flutter and vibration tests. The airfoil may undergo vertical and torsional (about a spanwise axis) displacements (h and α, respectively), restrained only by rectilinear and torsion springs as shown in Fig. 2-12. Write the equations of motion for the case of a nonoperating tunnel, i.e., zero airspeed.

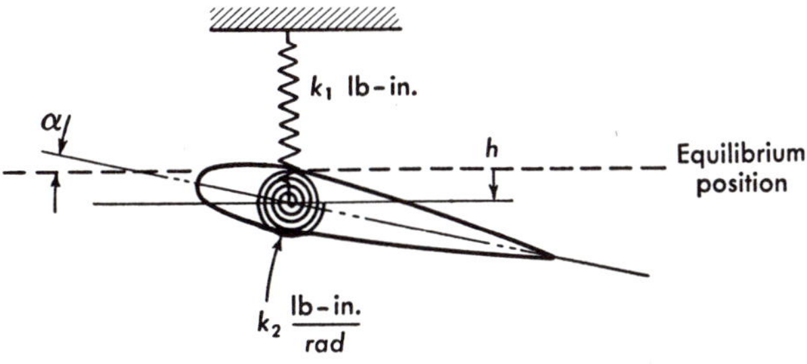

FIG. 2-12. Problem 8.

2-9. In the system of Fig. 2-2, assume $k_1 = k_2 = k_3 = k$ and $m_1 = m_2 = m$. Calculate the fundamental frequency by means of Rayleigh's method (see page 37) and compare with that obtained from the exact analysis, Eq. 2-17a.

2-10. Consider a cantilever beam of weight per unit length w and a concentrated weight W at the free end. Using Rayleigh's method, derive the expression for the fundamental frequency (Eq. 2-41).

2-11. An airplane wing is idealized to a cantilever beam of weight per unit length w. If the airload is distributed according to a linear variation of p lb/ft at the root and zero at the tip (see Fig. 2-13), calculate the fundamental frequency by Rayleigh's method. Let $p = 3w$.

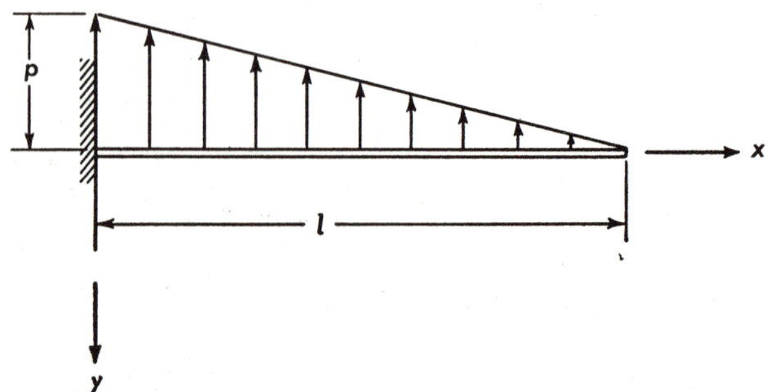

FIG. 2–13. Problem 11.

2–12. Find a general solution for a vibrating, uniform cantilever beam by expressing the deflection shape in the form of a trigonometric series.

2–13. From the frequency equation for a uniform cantilever beam,

$$1 + \cos kl \cosh kl = 0$$

determine the numerical values of the first three principal frequencies. Sketch the first three mode shapes.

2–14. Using the general solution (Eq. 2–61) of the vibrating beam problem, determine the frequency equation for a simply supported beam. Sketch the first three mode shapes and calculate the corresponding frequencies.

2–15. Using the general solution (Eq. 2–61) of the vibrating beam problem, determine the frequency equation for a beam which is free at both ends. Discuss the various types of mode shapes which such a beam would exhibit.

REFERENCES

1. TIMOSHENKO, S. P. *Vibration Problems in Engineering.* 3d ed.; New York: D. Van Nostrand Co., Inc., 1955.
2. DEN HARTOG, J. P. *Mechanical Vibrations.* 3d ed.; New York: McGraw-Hill Book Co., Inc., 1947.
3. MORRILL, B., *Mechanical Vibrations.* New York: The Ronald Press Co., 1957.
4. VON KÁRMÁN, T., and BIOT, M. A. *Mathematical Methods in Engineering.* New York: McGraw-Hill Book Co., Inc., 1940. Ch. III.
5. SLATER, J. C., and FRANK, N. H. *Mechanics.* New York: McGraw-Hill Book Co., Inc., 1947. Ch. IV.
6. RUTHERFORD, D. E. *Classical Mechanics.* New York: Interscience Publishers, Inc., 1951. Ch. V.
7. LAMB, H. *Dynamics.* New York: Cambridge University Press, 1923. Ch. XIII.
8. HOUBOLT, J. C., and ANDERSON, R. A. "Calculation of Uncoupled Modes and Frequencies in Bending or Torsion of Nonuniform Beams," *NACA Tech. Note* 1522 (February, 1948).
9. FETTIS, H. E. "A Modification of the Holzer Method for Computing Uncoupled Torsion and Bending Modes," *Jour. Aero. Sciences,* **16** (October, 1949); 625–634.
10. MYKLESTAD, N. O. *Vibration Analysis.* New York: McGraw-Hill Book Co., Inc., 1944. Ch. VI.
11. SCANLAN, R. H., and ROSENBAUM, R. *Aircraft Vibration and Flutter.* New York: The Macmillan Co., 1951. Ch. VII. *(Dover reprint)*

CHAPTER 3

ELEMENTS OF MATRIX ALGEBRA WITH APPLICATIONS

3–1. Introduction. As the reader may have already observed in previous chapters, vibration theory involves not only the solution of differential equations but also of algebraic equations. The solution of such equations is often facilitated by the use of determinants and matrices. Furthermore, problems requiring lengthy numerical computations may often be set up in a simpler manner by the use of matrix algebra. Because most dynamics problems encountered in aeronautical engineering involve the solutions of algebraic equations and require lengthy numerical computations, the use of the matrix algebra and notation has become widespread. It is worthwhile, therefore, to spend some effort in learning the new notation and conventions peculiar to this important mathematical tool.

3–2. Determinants. In the preceding chapter it was seen that the study of free vibrations of multi-degree-of-freedom systems always leads to the evaluation of a determinant. Therefore, it is worthwhile here to review a few of the fundamental properties of determinants not only as a prerequisite for the study of matrix algebra which follows, but also for their own importance in vibration analysis.

A *determinant* is a square array of quantities arranged so that a_{ij} represents the element in the ith row and j^{th} column. Thus, the determinant $|\,a\,|$ is written as

$$|\,a\,| = \begin{vmatrix} a_{11} & a_{12} & \cdots & a_{1n} \\ a_{21} & a_{22} & \cdots & a_{2n} \\ \cdot & & & \\ \cdot & & & \\ \cdot & & & \\ a_{n1} & a_{n2} & \cdots & a_{nn} \end{vmatrix} \tag{3–1}$$

Numerical values of second- and third-order determinants may be obtained by expansion along diagonals. For example, the numerical value of the determinant

$$\begin{vmatrix} a_{11} & a_{12} \\ a_{21} & a_{22} \end{vmatrix}$$

may be obtained from its expansion: $a_{11}a_{22} - a_{12}a_{21}$.

Determinants appear logically in the solution of linear algebraic equations since the solutions of the two equations

$$a_{11}x + a_{12}y = c_{11}$$

$$a_{21}x + a_{22}y = c_{21}$$

may be found in determinant form as

$$x = \frac{\begin{vmatrix} c_{11} & a_{12} \\ c_{21} & a_{22} \end{vmatrix}}{\begin{vmatrix} a_{11} & a_{12} \\ a_{21} & a_{22} \end{vmatrix}} = \frac{a_{22}c_{11} - a_{12}c_{21}}{a_{11}a_{22} - a_{21}a_{12}}$$

$$y = \frac{\begin{vmatrix} a_{11} & c_{11} \\ a_{21} & c_{21} \end{vmatrix}}{\begin{vmatrix} a_{11} & a_{12} \\ a_{21} & a_{22} \end{vmatrix}} = \frac{a_{11}c_{21} - a_{21}c_{11}}{a_{11}a_{22} - a_{21}a_{12}}$$

The *minor* of the element a_{ij} of the determinant (Eq. 3–1) is obtained by striking out the ith row and the jth column. Thus, in the determinant

$$\begin{vmatrix} a_{11} & a_{12} & a_{13} \\ a_{21} & a_{22} & a_{23} \\ a_{31} & a_{32} & a_{33} \end{vmatrix}$$

the minor of the element a_{21} is

$$M_{21} = \begin{vmatrix} a_{12} & a_{13} \\ a_{32} & a_{33} \end{vmatrix}$$

The *cofactor* of the element a_{ij} of the determinant (Eq. 3–1) is obtained from the minor of that element with a sign affixed to it according to the relation

$$C_{ij} = (-1)^{i+j}M_{ij} \tag{3–2}$$

Thus, in the third-order determinant above, the cofactor of element a_{21} is given by

$$C_{21} = (-1)^3 M_{ij} = -\begin{vmatrix} a_{12} & a_{13} \\ a_{32} & a_{33} \end{vmatrix}$$

A determinant of nth order may be conveniently reduced to a sum of determinants of the $(n - 1)^{th}$ order by means of the *Laplace expansion theorem.* The determinant $| a |$ of nth order may be written in terms of the cofactors of elements of the first column, or row, as

$$| a | = \sum_{j=1}^{n} a_{ij}C_{ij} = \sum_{i=1}^{n} a_{ij}C_{ij} \tag{3–3}$$

Thus, we may expand a fourth-order determinant as a sum of third-order determinants:

$$\begin{vmatrix} a_{11} & a_{12} & a_{13} & a_{14} \\ a_{21} & a_{22} & a_{23} & a_{24} \\ a_{31} & a_{32} & a_{33} & a_{34} \\ a_{41} & a_{42} & a_{43} & a_{44} \end{vmatrix} = a_{11}C_{11} + a_{21}C_{21} + a_{31}C_{31} + a_{41}C_{41}$$

$$= a_{11}\begin{vmatrix} a_{22} & a_{23} & a_{24} \\ a_{32} & a_{33} & a_{34} \\ a_{42} & a_{43} & a_{44} \end{vmatrix} - a_{21}\begin{vmatrix} a_{12} & a_{13} & a_{14} \\ a_{32} & a_{33} & a_{34} \\ a_{42} & a_{43} & a_{44} \end{vmatrix}$$

$$+ a_{31}\begin{vmatrix} a_{12} & a_{13} & a_{14} \\ a_{22} & a_{23} & a_{24} \\ a_{42} & a_{43} & a_{44} \end{vmatrix} - a_{41}\begin{vmatrix} a_{12} & a_{13} & a_{14} \\ a_{22} & a_{23} & a_{24} \\ a_{32} & a_{33} & a_{34} \end{vmatrix}$$

We see that repeated application of the Laplace expansion theorem enables us to evaluate almost any numerical determinant very easily.

The following elementary properties of determinants may be useful.

(a) If all elements of any row or column are zero, then the determinant is zero.

(b) If two rows or two columns of a determinant are identical, then the determinant is zero.

(c) If all the elements in a column are multiplied by a given factor, then the determinant is multiplied by that factor.

(d) If all the elements in a row or column except one are zero, then the determinant is equal to that element times its cofactor.

(e) The interchange of any two columns or rows changes the sign of the determinant.

Further properties of determinants and detailed expositions of methods of evaluating numerical determinants may be found in the references cited at the end of this chapter.

3–3. Matrices and Matrix Algebra. A *matrix* is an array of quantities in which a_{ij} indicates the element of the ith row and j^{th} column. A matrix is denoted by enclosure within square brackets. Thus the matrix $[a]$ is given by

$$[a] = \begin{bmatrix} a_{11} & a_{12} & \cdots & a_{1n} \\ a_{21} & a_{22} & \cdots & a_{2n} \\ \cdot & & & \\ \cdot & & & \\ \cdot & & & \\ a_{m1} & a_{m2} & \cdots & a_{mn} \end{bmatrix} \tag{3–4}$$

A *square* matrix is one in which the number of rows is equal to the number of columns.

A *unit* matrix is a square matrix in which the diagonal elements (a_{ii}) are all equal to unity and all other elements are zero (sometimes called the *identity* matrix).

A *diagonal* matrix is a square matrix in which all elements except the diagonal elements are zero.

A *row* matrix is one composed of only a single row (sometimes called a *vector*):

$$[b] = [b_{11}\, b_{12} \cdots b_{1n}] \tag{3–5a}$$

A *column* matrix is one composed of only a single column (sometimes called a *point*). A column matrix is usually denoted by braces in order to conserve space; thus

$$\{b\} = \begin{bmatrix} b_{11} \\ b_{21} \\ \cdot \\ \cdot \\ \cdot \\ b_{m1} \end{bmatrix} = \{b_{11}\, b_{21} \cdots b_{m1}\} \tag{3–5b}$$

If all the rows and columns of a matrix are completely interchanged, the new matrix is called a *transposed* matrix and is denoted by a prime. Thus, $[a]'$ is the transpose of $[a]$. The transpose of a row matrix is a column matrix and the transpose of a column matrix is a row matrix. If, in a given matrix, $a_{ij} = a_{ji}$, then $[a]' = [a]$ and the matrix $[a]$ is a *symmetrical* matrix.

Two matrices are equal if the corresponding elements in each are equal. Thus, $[a] = [b]$ if $a_{ij} = b_{ij}$.

Two matrices of the same order (same number of rows and columns) may be added to form a new matrix whose elements are sums of the elements of the two original matrices. Therefore,

$$[c] = [a] + [b], \quad c_{ij} = a_{ij} + b_{ij} \tag{3–6a}$$

Two matrices may be subtracted in a similar fashion:

$$[d] = [a] - [b], \quad d_{ij} = a_{ij} - b_{ij} \tag{3-6b}$$

The multiplication of two matrices is somewhat more complicated. In order to multiply two matrices together they must be *conformable;* that is, the number of columns of the first must be equal to the number of rows of the second. The product matrix, which has the same number of rows and columns as does the second matrix, may be best illustrated by means of an example.

$$[a][b] = \begin{bmatrix} a_{11} & a_{12} & a_{13} \\ a_{21} & a_{22} & a_{23} \\ a_{31} & a_{32} & a_{33} \end{bmatrix} \begin{bmatrix} b_{11} & b_{12} \\ b_{21} & b_{22} \\ b_{31} & b_{32} \end{bmatrix}$$

$$= \begin{bmatrix} (a_{11}b_{11} + a_{12}b_{21} + a_{13}b_{31}) & (a_{11}b_{12} + a_{12}b_{22} + a_{13}b_{32}) \\ (a_{21}b_{11} + a_{22}b_{21} + a_{23}b_{31}) & (a_{21}b_{12} + a_{22}b_{22} + a_{23}b_{32}) \\ (a_{31}b_{11} + a_{32}b_{21} + a_{33}b_{31}) & (a_{31}b_{12} + a_{32}b_{22} + a_{33}b_{32}) \end{bmatrix}$$

Close study of this result will reveal that a row-by-column process has been used. For instance, the element in the first row and first column of the product matrix is obtained by "multiplying" the first row of $[a]$ into the first column of $[b]$ and adding the corresponding terms. All elements of the first column of the product matrix come from "multiplication" of the first column of the $[b]$ matrix by the rows of the $[a]$ matrix. Because $[a][b] \neq [b][a]$, we must be careful to note which product is to be obtained.

The rule for matrix multiplication may therefore be written concisely in terms of the elements c_{ij} of the product matrix as

$$c_{ij} = \sum_{k=1}^{p} a_{ik}b_{kj} \tag{3-7}$$

where p is the number of columns of the first matrix.

The process of matrix division is considerably different from ordinary algebraic division. The procedure is accomplished by means of an *inverse* matrix $[a]^{-1}$ defined in such a manner that

$$[a]^{-1}[a] = [U]$$

where $[U]$ is the unit matrix. Thus, if the quotient $[c]$ of $[b]$ divided by $[a]$ is given by

$$[a][c] = [b]$$

we may multiply both sides by the inverse of $[a]$ and obtain

$$[a]^{-1}[a][c] = [U][c] = [a]^{-1}[b]$$

The inverse of the matrix $[a]$ may be found as the quotient of the transposed matrix, whose elements are the cofactors of $[a]$, and the determinant of $[a]$. Thus, since

$$[C_{ij}]' = [C_{ji}]$$

we have

$$[a]^{-1} = \frac{[C_{ji}]}{|a|} \tag{3–8}$$

We thus require that $|a| \neq 0$. The matrix $[C_{ji}]$ is called the *adjoint* of $[a]$.

For example, let us find the inverse of the second order matrix

$$[a] = \begin{bmatrix} a_{11} & a_{12} \\ a_{21} & a_{22} \end{bmatrix}$$

The transpose of $[a]$ is

$$[a]' = \begin{bmatrix} a_{11} & a_{21} \\ a_{12} & a_{22} \end{bmatrix}$$

The adjoint of $[a]$ is

$$[C_{ji}] = \begin{bmatrix} a_{22} & -a_{12} \\ -a_{21} & a_{11} \end{bmatrix}$$

The determinant of $[a]$ is $a_{11}a_{22} - a_{12}a_{21}$. Therefore, the inverse of $[a]$ is

$$[a]^{-1} = \frac{1}{a_{11}a_{22} - a_{12}a_{21}} \begin{bmatrix} a_{22} & -a_{12} \\ -a_{21} & a_{11} \end{bmatrix}$$

Let us now solve the two algebraic equations treated earlier

$$a_{11}x + a_{12}y = c_{11}$$

$$a_{21}x + a_{22}y = c_{21}$$

by matrix algebra. We introduce the matrices

$$[a] = \begin{bmatrix} a_{11} & a_{12} \\ a_{21} & a_{22} \end{bmatrix}$$

$$\{z\} = \{x \quad y\}$$

$$\{c\} = \{c_{11} \quad c_{21}\}$$

The two algebraic equations may then be written in matrix notation as

$$[a]\{z\} = \{c\}$$

Upon solving for $\{z\}$ we have

$$[a]^{-1}[a]\{z\} = [a]^{-1}\{c\}$$
$$\{z\} = [a]^{-1}\{c\}$$

But, from the preceding example,

$$[a]^{-1} = \frac{1}{|a|}\begin{bmatrix} a_{22} & -a_{12} \\ -a_{21} & a_{11} \end{bmatrix}$$

so that

$$\begin{bmatrix} x \\ y \end{bmatrix} = \frac{1}{|a|}\begin{bmatrix} a_{22} & -a_{12} \\ -a_{21} & a_{11} \end{bmatrix}\begin{bmatrix} c_{11} \\ c_{21} \end{bmatrix}$$
$$= \frac{1}{|a|}\begin{bmatrix} a_{22}c_{11} & -a_{12}c_{21} \\ -a_{21}c_{11} & +a_{11}c_{21} \end{bmatrix}$$

These two matrices are equal, so that we may equate corresponding elements; therefore,

$$x = \frac{a_{22}c_{11} - a_{12}c_{21}}{a_{11}a_{22} - a_{12}a_{21}}$$

$$y = \frac{a_{11}c_{21} - a_{21}c_{11}}{a_{11}a_{22} - a_{12}a_{21}}$$

as before.

Positive and negative powers of square matrices are obtained easily by application of matrix multiplication and division. Thus,

$$[a]^n = [a]\cdot[a] \cdots \quad (n \text{ times}) \tag{3–9a}$$

and

$$([a])^{-n} = ([a]^{-1})^n \tag{3–9b}$$

The product of two matrices $[a]$ and $[b]$ is a new matrix $[c]$ whose elements are given by Eq. 3–7. When transposed, however, $[a]$ and $[b]$ will be conformable when multiplied as $[b]\cdot[a]$, resulting in a matrix whose elements are

$$c_{ji} = \sum_{k=1}^{p} b_{ik}a_{kj}$$

so that when a matrix product is transposed the order of the matrices must be reversed. Thus,

$$([a][b])' = [b]'[a]' \tag{3–10}$$

and similarly

$$([a][b][c])' = [c]'[b]'[a]' \tag{3-11}$$

A similar result holds in the case of the inverse of a matrix product. Thus, if

$$[c] = [a][b]$$

then

$$[b]^{-1}[a]^{-1}[c] = [b]^{-1}[a]^{-1}[a][b]$$

from which

$$[b]^{-1}[a]^{-1} = [c]^{-1} = ([a][b])^{-1} \tag{3-12}$$

Because of the simple and general definition of a matrix, it is not uncommon that the elements of a matrix are complex numbers or even matrices themselves. In the former case, a number of special matrices are often defined as follows:

(a) Conjugate matrix: the matrix $[\bar{a}]$ whose elements are complex conjugates of those of the matrix $[a]$.
(b) Hermitean matrix: $[a]$ is Hermitean if $[a] = [\bar{a}]'$.
(c) Associate matrix: $[\bar{a}]'$ is the associate matrix of $[a]$.
(d) Unitary matrix: $[a]$ is unitary if $[a] = ([\bar{a}]')^{-1}$.

Other matrices of similar nature have also been defined; we shall not give them here.

In many instances a matrix may be conveniently thought of as having elements which are themselves matrices. For example,

$$[a] = \begin{bmatrix} 1 & 2 & 7 & 6 \\ 4 & 3 & 6 & 5 \\ 2 & 0 & 1 & 2 \\ 3 & 5 & 2 & 1 \end{bmatrix}$$

can be written as

$$[a] = \begin{bmatrix} A & B \\ C & D \end{bmatrix}$$

where

$$[A] = \begin{bmatrix} 1 & 2 \\ 4 & 3 \end{bmatrix}, \quad [B] = \begin{bmatrix} 7 & 6 \\ 6 & 5 \end{bmatrix}$$

$$[C] = \begin{bmatrix} 2 & 0 \\ 3 & 5 \end{bmatrix}, \quad [D] = \begin{bmatrix} 1 & 2 \\ 2 & 1 \end{bmatrix}$$

or as

$$[a] = \begin{bmatrix} E & F \\ G & H \end{bmatrix}$$

where

$$[E] = \begin{bmatrix} 1 & 2 & 7 \\ 4 & 3 & 6 \\ 2 & 0 & 1 \end{bmatrix}, \quad [F] = \{6 \quad 5 \quad 2\}$$

$$[G] = [\,3 \quad 5 \quad 2\,], \qquad [H] = [1]$$

or in a variety of other ways. Writing the matrix $[a]$ in any of these ways is called "partitioning." Partitioning is usually denoted by dashed lines in the original matrix. Thus, in the second example preceding,

$$[a] = \left[\begin{array}{ccc:c} 1 & 2 & 7 & 6 \\ 4 & 3 & 6 & 5 \\ 2 & 0 & 1 & 2 \\ \hdashline 3 & 5 & 2 & 1 \end{array}\right]$$

The algebra of matrices presented earlier applies equally well to matrices with complex elements or to partitioned matrices.

Let $[x(t)]$ be a matrix whose elements x_{ij} are functions of a variable t:

$$[x(t)] = \begin{bmatrix} x_{11}(t) & x_{12}(t) & \cdots & x_{1n}(t) \\ x_{21}(t) & x_{22}(t) & \cdots & x_{2n}(t) \\ \cdot & \cdot & \cdot & \cdot \\ x_{m1}(t) & x_{m2}(t) & \cdots & x_{mn}(t) \end{bmatrix} \tag{3–13}$$

Differentiation and integration of $[x(t)]$ is performed by operating on the individual elements. Thus,

$$\frac{d}{dt}[x(t)] = \begin{bmatrix} \dfrac{dx_{11}}{dt} & \dfrac{dx_{12}}{dt} & \cdots & \dfrac{dx_{1n}}{dt} \\ \dfrac{dx_{21}}{dt} & \dfrac{dx_{22}}{dt} & \cdots & \dfrac{dx_{2n}}{dt} \\ \cdot & \cdot & \cdot & \cdot \\ \dfrac{dx_{m1}}{dt} & \dfrac{dx_{m2}}{dt} & \cdots & \dfrac{dx_{mn}}{dt} \end{bmatrix} \tag{3–14}$$

and

$$\int [x(t)]\,dt = \begin{bmatrix} \int x_{11}\,dt & \int x_{12}\,dt & \cdots & \int x_{1n}\,dt \\ \int x_{21}\,dt & \int x_{22}\,dt & \cdots & \int x_{2n}\,dt \\ \cdot & \cdot & \cdot & \cdot \\ \int x_{m1}\,dt & \int x_{m2}\,dt & \cdots & \int x_{mn}\,dt \end{bmatrix} \tag{3-15}$$

3-4. Application of Matrices to the Theory of Equations and Characteristic-Value Problems. In the study of systems of linear equations, the concept of linear independence of quantities is of importance.

A set of quantities $f_1, f_2, \cdots f_m$ $(m \geq 2)$ is said to be linearly independent if there do not exist m constants $c_1, c_2, \cdots c_m$, which are not all zero, such that

$$c_1 f_1 + c_2 f_2 + \cdots + c_m f_m = 0$$

If the set is linearly dependent, then at least one of the f_i can be expressed as a linear combination of the others.

A test for linear independence may be given in terms of the *rank* of a certain matrix. If $[A]$ is a matrix, and if all minors of $[A]$ of order $r + 1$ are zero while at least one minor of order r is not zero, the matrix $[A]$ is said to be of rank r. The test for linear independence is as follows: The linear functions $f_i = a_{i1}x_1 + a_{i2}x_2 + \cdots + a_{in}x_n$ $(i = 1, 2, \cdots m)$ are linearly independent if, and only if, the matrix of the coefficients $[a]$ is of rank $r \geq m$. Any set of linear functions in less than m unknowns must be linearly dependent.

A set of equations that have at least one common solution is said to be a *consistent set* of equations. A set for which there exists no common solution is called an inconsistent set.

Consider the set of linear nonhomogeneous equations

$$\begin{aligned} a_{11}x_1 + a_{12}x_2 + \cdots + a_{1n}x_n &= k_1 \\ a_{21}x_1 + a_{22}x_2 + \cdots + a_{2n}x_n &= k_2 \\ \cdot \quad \cdot \quad \cdot \quad \cdot \quad \cdot \quad & \quad \cdot \\ a_{m1}x_1 + a_{m2}x_2 + \cdots + a_{mn}x_n &= k_m \end{aligned} \tag{3-16}$$

where at least one $k_i \neq 0$. If the matrix of the coefficients $[a]$ is of rank r, the set of equations is consistent, provided that the rank of the matrix

$$[k] = \begin{bmatrix} a_{11} & a_{12} & \cdots & a_{1n} & k_1 \\ a_{21} & a_{22} & \cdots & a_{2n} & k_2 \\ \cdot & \cdot & \cdot & \cdot & \cdot \\ a_{m1} & a_{m2} & \cdots & a_{mn} & k_m \end{bmatrix} \tag{3-17}$$

which is called the *augmented matrix*, is also r.

If the set of equations above is homogeneous, then it will have a solution different from the trivial one, $x_1 = x_2 = \cdots x_n = 0$, if the rank of the coefficient matrix $[a]$ is less than n. In the case of $m = n$, this is equivalent to the statement that in order for solutions other than the trivial one to exist, the determinant of the coefficients must vanish, inasmuch as the rank of the coefficient matrix will then be less than n.

Frequently, it is desired to transform a set of equations from one variable to another by means of a linear process. Let us suppose that y is a prescribed function of u and x and that z is a prescribed function of v and y. It is desired to express x as a function of z. Suppose that the prescribed functions are

$$\begin{aligned} u_{11}x_1 + u_{12}x_2 + \cdots + u_{1n}x_n &= y_1 \\ u_{21}x_1 + u_{22}x_2 + \cdots + u_{2n}x_n &= y_2 \\ \cdot \quad \cdot \quad \cdot \quad \cdot \quad \cdot \quad \cdot \quad \cdot &\quad \cdot \quad \cdot \\ u_{n1}x_1 + u_{n2}x_2 + \cdots + u_{nn}x_n &= y_n \end{aligned} \tag{3–18a}$$

and

$$\begin{aligned} v_{11}y_1 + v_{12}y_2 + \cdots + v_{1n}y_n &= z_1 \\ v_{21}y_1 + v_{22}y_2 + \cdots + v_{2n}y_n &= z_2 \\ \cdot \quad \cdot \quad \cdot \quad \cdot \quad \cdot \quad \cdot \quad \cdot &\quad \cdot \quad \cdot \\ v_{n1}y_1 + v_{n2}y_2 + \cdots + v_{nn}y_n &= z_n \end{aligned} \tag{3–19a}$$

These two sets of equations may be written in matrix form as

$$[u]\{x\} = \{y\} \tag{3–18b}$$

$$[v]\{y\} = \{z\} \tag{3–19b}$$

where the matrices $[u]$ and $[v]$ are called "transformation" matrices. We may combine these two equations to the form of z as a function of x by direct substitution:

$$\{z\} = [v][u]\{x\} \tag{3–20}$$

If $[u]$ and $[v]$ are both nonsingular, we may also write x as a function of z:

$$\{x\} = [u]^{-1}[v]^{-1}\{z\} \tag{3–21}$$

The solution of a set of n linear equations in n unknowns is easily accomplished by means of the matrix algebra. For example, the set of equations

$$a_{i1}x_1 + a_{i2}x_2 + \cdots + a_{in}x_n = k_i, \quad (i = 1, 2, \cdots n) \tag{3–22a}$$

may be written in the matrix form

$$[a]\{x\} = \{k\} \tag{3–22b}$$

Premultiplying by the inverse of $[a]$, assumed to be nonsingular, the solution is given as

$$\{x\} = [a]^{-1}\{k\} \tag{3-23}$$

A frequently occurring problem in the applications is that of determining the values of a constant λ in the set of equations

$$a_{i1}x_1 + a_{i2}x_2 + \cdots + a_{in}x_n = \lambda x_i, \quad (i = 1, 2, \cdots n) \tag{3-24a}$$

for which nontrivial solutions exist. Such a problem is called a *characteristic-value problem* and the values of λ for which solutions, other than $x_i = 0$, exist are called *characteristic values* (sometimes "eigenvalues" or "latent roots") of the coefficient matrix $[a]$. The corresponding solutions for the elements of the column matrix $\{x\}$ are called the *characteristic vectors* (sometimes "eigenvectors") of the problem. The column matrix $\{x\}$ is often called the *modal column.*

In matrix form, the set of equations above is

$$[a]\{x\} = \lambda\{x\} \tag{3-24b}$$

or

$$([a] - \lambda[U])\{x\} = \{0\} \tag{3-25}$$

The matrix $([a] - \lambda[U])$ is called the *characteristic matrix,* and the determinant of this matrix $|\,[a] - \lambda[U]\,|$ is called the *determinantal equation* (sometimes the "characteristic" equation). The roots $\lambda_1, \lambda_2, \cdots \lambda_n$ of the determinantal equation are then the characteristic values defined above.

An important theorem related to the characteristic-value problem, and known as the *Cayley-Hamilton theorem,* may be stated as follows: Let

$$f(\lambda) = (-1)^n\left(\lambda^n - p_1\lambda^{n-1} + \cdots + (-1)^n p_n\right) = 0$$

be the characteristic equation of a square matrix $[a]$ of order n. Then the matrix polynomial equation

$$F([X]) = [X]^n - p_1[X]^{n-1} + \cdots + (-1)^n p_n[U] = [0]$$

is satisfied by $[X] = [a]$.

This theorem is often stated as: "a matrix satisfies its own characteristic equation."

To prove the theorem, let $[A]$ be the adjoint of the characteristic matrix $([a] - \lambda[U])$. Since the elements of $([a] - \lambda[U])$ are at most of the first degree in λ, their cofactors in $|\,[a] - \lambda[U]\,|$ are at most of degree $(n - 1)$ in λ. Hence,

$$[A] = [A_{n-1}]\lambda^{n-1} + [A_{n-2}]\lambda^{n-2} + \cdots + [A_0]$$

We may also write

$$f(\lambda) = k_n\lambda^n + k_{n-1}\lambda^{n-1} + \cdots + k_0$$

By using the formula

$$[a][A_{ji}] = |a|[U]$$

we see that

$$[a][A] - \lambda[A] \equiv f(\lambda)[U]$$

Substituting the previous expressions for $[A]$ and $f(\lambda)$ into this relation, we have, upon equating corresponding powers of λ,

$$\begin{aligned} [a][A_0] &= k_0[U] \\ [a][A_1] - [A_0] &= k_1[U] \\ [a][A_2] - [A_1] &= k_2[U] \\ \cdots\cdots&\cdots\cdots \\ [a][A_{n-1}] - [A_{n-2}] &= k_{n-1}[U] \\ -[A_{n-1}] &= k_n[U] \end{aligned}$$

If we multiply these equations in succession by $[U]$, $[a]$, $[a]^2$, $\cdots$ $[a]^n$, and add, the first numbers cancel out, and we get

$$k_0[U] + k_1[a] + k_2[a]^2 + \cdots + k_n[a]^n = [0]$$

or

$$f([a]) = [0]$$

which is the statement of the theorem.

In almost all physical applications arising in airplane dynamics the matrix $[a]$ is symmetrical; therefore the following discussions will be restricted to that case. We shall also restrict ourselves to the case of all λ_i being distinct, as is true in physical problems.

An interesting and useful property of the characteristic vectors will now be demonstrated. Let λ_p and λ_q be distinct latent roots of $[a]$. For the equation

$$([a] - \lambda[U])\{x\} = \{0\} \tag{3–25}$$

the root $\lambda = \lambda_p$ gives

$$[a]\{x_p\} = \lambda_p\{x_p\} \tag{3–26}$$

while for the analogous equation

$$\{x\}'([a] - \lambda[U]) = [0] \tag{3–27}$$

where $\{x\}'$ is a row matrix, the root $\lambda = \lambda_q$ gives

$$\{x_q\}'[a] = \lambda_q\{x_q\}' \tag{3-28}$$

Postmultiplying by $[a]^{-1}$ gives

$$\{x_q\}' = \lambda_q\{x_q\}'[a]^{-1}$$

and postmultiplying again by $[a]\{x_p\}$ gives

$$\{x_q\}'[a]\{x_p\} = \lambda_q\{x_q\}'\{x_p\}$$

which may be rewritten as

$$\lambda_p\{x_q\}'\{x_p\} = \lambda_q\{x_q\}'\{x_p\}$$

Thus

$$(\lambda_p - \lambda_q)\{x_q\}'\{x_p\} = \{0\}$$

and since $\lambda_p \neq \lambda_q$, we have

$$\{x_q\}'\{x_p\} = \{0\}, \quad (p \neq q) \tag{3-29}$$

This result is known as the generalized *orthogonality property* of characteristic rows and columns. Some additional comments relating to orthogonality conditions are given in Appendix B.

In practice, the necessity for expanding the determinantal equation and solving the resulting high-degree polynomial equation is a serious obstacle to numerical evaluation of the latent roots. However, there exists a numerical iterative method which allows the evaluation of the latent roots, and the modal columns as well, without actual expansion of the determinantal equation.

From the equation

$$([a] - \lambda[U])\{x\} = \{0\} \tag{3-25}$$

we see that the rth modal column satisfies the equation

$$[a]\{x^{(r)}\} = \lambda_r\{x^{(r)}\} \tag{3-30}$$

Postmultiplying by $[a]$, p times, we have

$$[a]^p\{x^{(r)}\} = \lambda_r^p\{x^{(r)}\}$$

We have n equations of this type, one for each modal column $\{x^{(r)}\}$ and its associated latent root λ_r. Let us now construct a square matrix $[X]$ from the modal columns $\{x^{(r)}\}$ in the following manner:

$$[X] = [\{x^{(1)}\} \quad \{x^{(2)}\} \quad \cdots \quad \{x^{(n)}\}]$$

The set of equations above may now be written in the form

$$[a]^p[X] = [X]\begin{bmatrix} \lambda_1{}^p & 0 & 0 & \cdots & 0 \\ 0 & \lambda_2{}^p & 0 & \cdots & 0 \\ \cdot & \cdot & \cdot & \cdot & \cdot \\ 0 & 0 & 0 & \cdots & \lambda_n{}^p \end{bmatrix} \tag{3-31}$$

Postmultiplying by $[X]^{-1}$, this becomes

$$[a]^p = [X][\lambda][X]^{-1} \tag{3-32}$$

Let the roots of the determinantal equation be so arranged that $\lambda_1 > \lambda_2 > \cdots \lambda_n$. For p very large in the above equation, $\lambda_1 \gg \lambda_2$; thus, we may neglect all terms except those corresponding to the dominant root λ_1. Then, by direct multiplication,

$$\lim_{p\to\infty} [a]^p = \lambda_1{}^p \begin{bmatrix} X_{11}X_{11}{}^{-1} & X_{11}X_{12}{}^{-1} & \cdots & X_{11}X_{1n}{}^{-1} \\ X_{21}X_{11}{}^{-1} & X_{21}X_{12}{}^{-1} & \cdots & X_{21}X_{1n}{}^{-1} \\ \cdot & \cdot & \cdot & \cdot \\ X_{n1}X_{11}{}^{-1} & X_{n1}X_{12}{}^{-1} & \cdots & X_{n1}X_{1n}{}^{-1} \end{bmatrix} \tag{3-33}$$

This property of the $[a]$ matrix proves to be very useful in that it allows the development of a procedure that will yield the dominant root λ_1, as well as the modal column $\{x^{(1)}\}$.

We select an *arbitrary* column matrix $\{x\}_0$ and form the following sequence:

$$\begin{aligned} [a]\{x\}_0 &= \{x\}_1 \\ [a]\{x\}_1 &= [a]^2\{x\}_0 = \{x\}_2 \\ [a]\{x\}_2 &= [a]^3\{x\}_0 = \{x\}_3 \\ &\cdots\cdots\cdots \\ [a]\{x\}_{p-1} &= [a]^p\{x\}_0 = \{x\}_p \end{aligned}$$

Thus, in view of the foregoing result, for sufficiently large p,

$$\{x\}_p = [a]^p\{x\}_0 = \lambda_1{}^p \begin{bmatrix} X_{11}(X_{11}{}^{-1}x_{10} + X_{12}x_{20} + \cdots + X_{1n}x_{n0}) \\ X_{21}(X_{11}{}^{-1}x_{10} + X_{12}x_{20} + \cdots + X_{1n}x_{n0}) \\ \cdot \\ \cdot \\ \cdot \\ X_{n1}(X_{11}{}^{-1}x_{10} + X_{12}x_{20} + \cdots + X_{1n}x_{n0}) \end{bmatrix} \tag{3-34}$$

That is, by repeated multiplication of the matrix $[a]$ by the arbitrary column matrix $\{x\}_0$, we eventually reach a point where further multiplication of $[a]$ merely multiplies every element of the column matrix $\{x\}_{p-1}$ by a common factor. This common factor is the dominant root

λ_1, and the elements of the column matrix $\{x\}_p$ are proportional to those of the modal matrix $\{X^{(1)}\}$.

We now turn to a procedure by which the other latent roots may be determined. Suppose that λ_1 is known exactly. Then, for a symmetrical matrix, all other characteristic vectors may be considered as orthogonal to $\{x^{(1)}\}$, as we have already seen. Hence, if we impose the orthogonality constraint, the resultant problem will possess those characteristic numbers and vectors which are in addition to λ_1 and $\{X^{(1)}\}$. But the orthogonality condition permits one of the components (say, x_r) to be expressed as a linear combination of the others. Hence, we may eliminate x_r from the scalar equations corresponding to the characteristic matrix, thereby obtaining a set of $n - 1$ equations involving $n - 1$ components. The new dominant root and the corresponding characteristic vector are then obtained as before.

3-5. Application of Matrix Algebra to Vibration Problems. The application of the previous discussions to the physical problem of small oscillations of conservative systems will serve to illustrate many of the important points.

Suppose that we have a system of n degrees-of-freedom undergoing free undamped vibrations of small amplitude. The general equations of motion may be written as (see Art. 2-5)

$$\begin{aligned} m_{11}\ddot{q}_1 + m_{12}\ddot{q}_2 + \cdots + m_{1n}\ddot{q}_n + k_{11}q_1 + \cdots + k_{1n}q_n &= 0 \\ \cdots\cdots\cdots\cdots\cdots\cdots\cdots\cdots\cdots & \\ m_{n1}\ddot{q}_1 + m_{n2}\ddot{q}_2 + \cdots + m_{nn}\ddot{q}_n + k_{n1}q_1 + \cdots + k_{nn}q_n &= 0 \end{aligned} \tag{3-35}$$

We define the following matrices:

$$[M] = \begin{bmatrix} m_{11} & m_{12} & \cdots & m_{1n} \\ m_{21} & m_{22} & \cdots & m_{2n} \\ \cdot & & & \\ \cdot & & & \\ \cdot & & & \\ m_{n1} & m_{n2} & \cdots & m_{nn} \end{bmatrix} \equiv \text{inertia matrix} \tag{3-36}$$

$$[K] \equiv \begin{bmatrix} k_{11} & k_{12} & \cdots & k_{1n} \\ k_{21} & k_{22} & \cdots & k_{2n} \\ \cdot & & & \\ \cdot & & & \\ \cdot & & & \\ k_{n1} & k_{n2} & \cdots & k_{nn} \end{bmatrix} \equiv \text{stiffness matrix} \tag{3-37}$$

$$\{q\} \equiv \{q_1\, q_2 \cdots q_n\} \equiv \text{generalized coordinate matrix} \tag{3-38}$$

The equations of motion (Eq. 3–35) may now be written concisely in the matrix notation as

$$[M]\{\ddot{q}\} + [K]\{q\} = \{0\} \tag{3–39}$$

We assume solutions of the n equations (Eq. 3–39) of the form (see Art. 2–2 for the two-degree-of-freedom system):

$$\{q\} = \{A\} \sin(\omega t + \alpha) \tag{3–40}$$

where $\{A\}$ is a column matrix of arbitrary constants describing the amplitudes, and α is the phase angle. Introducing the assumed solutions (Eq. 3–40) into the equations of motion (Eq. 3–39), we have

$$([K] - \omega^2[M])\{A\} = \{0\} \tag{3–41}$$

These equations represent the algebraic equations for the unknown amplitudes (compare with Eq. 2–15). For convenience, multiply both sides of Eq. 3–41 by $[K]^{-1}$, which gives

$$([U] - \omega^2[D])\{A\} = \{0\} \tag{3–42}$$

where

$$[K]^{-1}[M] \equiv [D] \equiv \text{dynamical matrix} \tag{3–43}$$

The matrix $[K]^{-1}$ is often called the *flexibility* matrix, denoted by $[\Phi]$.

In order for solutions other than $\{A\} = \{0\}$ to exist, the determinant of the coefficients of $\{A\}$ must vanish (see Art. 2–2). Therefore,

$$\left| \frac{1}{\omega^2}[U] - [D] \right| = 0 \tag{3–44}$$

The solutions of this frequency equation give us the angular frequencies (principal frequencies) of the system. For each one of these real positive frequencies, ω_r, there corresponds a solution of the equations of motion of the form

$$\{q_r\} = \{A^{(r)}\} \sin(\omega_r t + \alpha_r) \tag{3–45}$$

Each of these equations represents a principal mode of vibration, and hence the column matrix $\{A\}$ is called a *modal column*. The general solution is made up of the sum of the principal modes; therefore,

$$\{q\} = \sum_{r=1}^{n} \{A^{(r)}\} \sin(\omega_r t + \alpha_r) \tag{3–46}$$

It should be emphasized again that we are not able to solve for the values of the amplitude constants A explicitly but only for their ratios.

Each modal column $\{A^{(r)}\}$ satisfies the characteristic matrix, with its corresponding value of ω_r, so that

$$\omega_r^2[M]\{A^{(r)}\} = [K]\{A^{(r)}\} \tag{3–47}$$

or

$$\{A^{(r)}\} = \frac{1}{\omega_r^2}\{A^{(r)}\}[D]^{-1} \tag{3-48}$$

We have seen previously (page 62) that this leads to the condition

$$\{A^{(r)}\}'[D]\{A^{(s)}\} = \lambda_r\{A^{(r)}\}'\{A^{(s)}\}$$

from which

$$\{A^{(r)}\}'\{A^{(s)}\} = \{0\}, \quad r \neq s \tag{3-49}$$

so that the principal modes of oscillation are orthogonal.

The frequency and mode shape of the lowest mode may be obtained by straightforward application of the iteration procedure given in the preceding article. The calculation of higher frequencies and mode shapes, however, requires additional comment.

It is not difficult to show that a typical row of the inverse of the modal matrix, that is, a row of $[A]^{-1}$, is proportional to the matrix product $\{A^{(r)}\}'[M]$. To do this, we write Eq. 3–31, for the case of $p = 1$, as

$$[A]^{-1}[D] = \begin{bmatrix} \lambda_1 & 0 & \cdots & 0 \\ 0 & \lambda_2 & \cdots & 0 \\ \cdot & \cdot & \cdot & \cdot \\ 0 & 0 & \cdots & \lambda_n \end{bmatrix}[A]^{-1}, \quad \lambda = \frac{1}{\omega^2} \tag{3-50}$$

If we take a typical row of $[A]^{-1}$ as

$$[A_r]^{-1} = [A_{r1}^{-1} \quad A_{r2}^{-1} \quad \cdots \quad A_{rn}^{-1}]$$

then by the above equation

$$[A_r]^{-1}[D] = \lambda_r[A_r]^{-1} \tag{3-51}$$

We also have, by the equation following Eq. 3–30, that

$$[D]\{A^{(r)}\} = \lambda_r\{A^{(r)}\}$$

or

$$[K]^{-1}[M]\{A^{(r)}\} = \lambda_r\{A^{(r)}\} \tag{3-52}$$

which can then be written as

$$\{A^{(r)}\}'[M][D] = \lambda_r\{A^{(r)}\}'[M] \tag{3-53}$$

Comparing this result with the equation above in terms of $[A_r]^{-1}$, we have

$$[A_r]^{-1} = \{A^{(r)}\}'[M] \tag{3-54}$$

This result then leads to an orthogonality condition for the row matrices $[A_r]^{-1}$ and the modal column matrices $\{A^{(r)}\}$ of the form

$$[A_r]^{-1}\{A^{(s)}\} = \{0\}, \quad r \neq s \tag{3-55}$$

If we premultiply the general solution by the row matrix $[A_1]^{-1}$, we have

$$[A_1]^{-1}\{q\} = C_1 \sin(\omega_1 t + \alpha_1) \tag{3-56}$$

If now the fundamental mode is to be absent, then the above orthogonality condition is employed, so that

$$[A_1]^{-1}\{q\} = \{0\} \tag{3-57}$$

Expanding this equation gives

$$A_{11}^{-1}q_1 + A_{12}^{-1}q_2 + \cdots + A_{1n}^{-1}q_n = 0 \tag{3-58}$$

from which we obtain equations of constraint between the coordinates in the form

$$\begin{aligned} q_1 &= -\frac{A_{12}^{-1}}{A_{11}^{-1}}q_2 - \frac{A_{13}^{-1}}{A_{11}^{-1}}q_3 \cdots - \frac{A_{1n}^{-1}}{A_{11}^{-1}}q_n \\ q_2 &= q_2 \\ &\cdot \quad \cdot \quad \cdot \\ q_n &= q_n \end{aligned} \tag{3-59}$$

This set of equations may be written in the form

$$\{q\} = \begin{bmatrix} 0 & -\dfrac{A_{12}^{-1}}{A_{11}^{-1}} & -\dfrac{A_{13}^{-1}}{A_{11}^{-1}} & \cdots & -\dfrac{A_{1n}^{-1}}{A_{11}^{-1}} \\ 0 & 1 & 0 & \cdots & 0 \\ \cdot & \cdot & \cdot & \cdot & \cdot \\ 0 & 0 & 0 & \cdots & 1 \end{bmatrix} \begin{bmatrix} q_1 \\ q_2 \\ \vdots \\ q_n \end{bmatrix} \tag{3-60}$$

or

$$\{q\} = [S]\{q\} \tag{3-61}$$

Now, the product of the matrix S with the original dynamic matrix $[D]$ gives a new dynamic matrix

$$[D][S] = [D]_2 \tag{3-62}$$

which is so constructed as to have the fundamental mode removed. Thus the iteration procedure applied to the matrix $[D]_2$ will yield the frequency and shape of the second mode.

3–6. A Numerical Example. The entire analysis is perhaps best illustrated by means of a simple example.* Let us consider the free vibrations of a double pendulum (Fig. 3–1). Let the small horizontal displacements of the masses, q_1 and q_2, be the generalized coordinates of

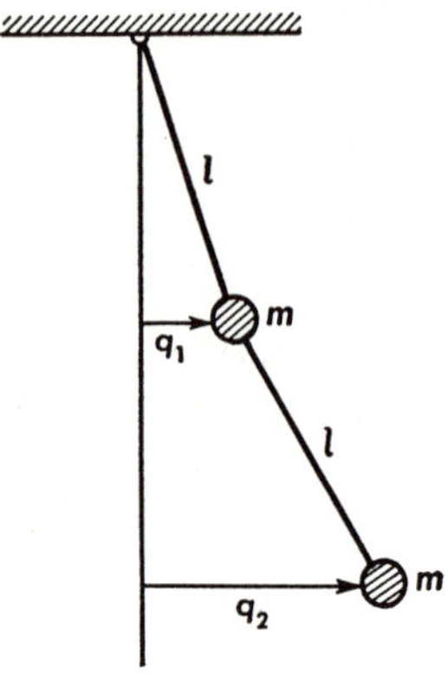

FIG. 3–1. Double pendulum.

the problem. If we apply a set of static forces F_1 and F_2 in the direction of q_1 and q_2, we may write the relation between the two as

$$\begin{bmatrix} F_1 \\ F_2 \end{bmatrix} = \begin{bmatrix} k_{11} & k_{12} \\ k_{21} & k_{22} \end{bmatrix} \begin{bmatrix} q_1 \\ q_2 \end{bmatrix} \tag{3–63}$$

where $[k]$ is the stiffness matrix. This may be written in terms of the flexibility matrix $[\Phi] \equiv [k]^{-1}$ as

$$\begin{bmatrix} q_1 \\ q_2 \end{bmatrix} = \begin{bmatrix} \varphi_{11} & \varphi_{12} \\ \varphi_{21} & \varphi_{22} \end{bmatrix} \begin{bmatrix} F_1 \\ F_2 \end{bmatrix} \tag{3–64}$$

The flexibility matrix is also known as the *matrix of influence coefficients* (see page 39). The elements of the flexibility matrix are obtained by applying unit loads and computing the resulting deflections. Thus, if a static unit force were applied horizontally to the first mass, the two masses would be displaced a distance $\bar{a}$ given by (taking moments about the suspension point)

$$\bar{a} = \frac{l}{2mg}$$

The first column of the flexibility matrix is thus $\varphi_{11} = \varphi_{21} = \bar{a}$. When a static unit force is applied to the lower mass, the displacement of the

* This example is identical with that worked out in References 1 and 3 except that the number of degrees of freedom has been reduced to two. Also see Problem 5, Chapter 2.

first mass will be $\bar{a}$, and the displacement of the lower mass will be

$$\bar{a} + \frac{l}{mg} = \bar{a} + \bar{b} = \frac{3}{2}\frac{l}{mg}$$

Thus, the second column of the flexibility matrix becomes $\varphi_{12} = \bar{a}$, $\varphi_{22} = \bar{a} + \bar{b}$. The complete flexibility matrix is then

$$[\Phi] = \begin{bmatrix} \bar{a} & \bar{a} \\ \bar{a} & \bar{a} + \bar{b} \end{bmatrix} \tag{3-65}$$

The mass (or inertia) matrix is very easily written as

$$[M] = \begin{bmatrix} m & 0 \\ 0 & m \end{bmatrix} \tag{3-66}$$

so that the dynamical matrix becomes

$$\begin{aligned} [D] = [\Phi][M] &= \begin{bmatrix} m\bar{a} & m\bar{a} \\ m\bar{a} & m(\bar{a} + \bar{b}) \end{bmatrix} \\ &= \frac{l}{2g}\begin{bmatrix} 1 & 1 \\ 1 & 3 \end{bmatrix} = \frac{l}{2g}[d] \end{aligned} \tag{3-67}$$

The determinantal equation may now be written (see Eq. 3–44) as

$$\left| \frac{2g}{l\omega^2}[U] - [d] \right| = \{0\} \tag{3-68}$$

To find the fundamental frequency, we assume an arbitrary column matrix

$$\{x\}_0 = \begin{bmatrix} 1 \\ 1 \end{bmatrix}$$

Then

$$[d]\begin{bmatrix} 1 \\ 1 \end{bmatrix} = \begin{bmatrix} 2 \\ 4 \end{bmatrix} = 4\begin{bmatrix} 0.5 \\ 1 \end{bmatrix}$$

Continuing the procedure, with $\{x\}_1 = \{0.5 \quad 1\}$,

$$[d]\begin{bmatrix} 0.5 \\ 1 \end{bmatrix} = \begin{bmatrix} 1.5 \\ 3.5 \end{bmatrix} = 3.5\begin{bmatrix} 0.428 \\ 1 \end{bmatrix}$$

$$[d]\begin{bmatrix} 0.43 \\ 1 \end{bmatrix} = \begin{bmatrix} 1.43 \\ 3.43 \end{bmatrix} = 3.43\begin{bmatrix} 0.417 \\ 1 \end{bmatrix}$$

Two more iterations will give the result

$$3.415\begin{bmatrix} 0.414 \\ 1 \end{bmatrix} = 3.415\{x\}_5$$

The fundamental frequency is found as

$$\frac{2g}{l\omega_1^2} = 3.415$$

or

$$\omega_1 = 0.765\sqrt{g/l} \tag{3-69}$$

The exact result is $\omega_1 = 0.768\sqrt{g/l}$. Thus, by taking a sufficient number of iterations and carrying a sufficient number of decimal places, we could easily obtain the value of the fundamental frequency to any desired degree of accuracy.

The modal column $\{A^{(1)}\}$ is proportional to the column $\{0.414 \quad 1\}$.

In order to determine the second mode shape and frequency, we follow the procedure outlined in the preceding article. The elements of the $[S]$ matrix are obtained from the modal column $\{A^{(1)}\}$ given above; thus,

$$[S] = \begin{bmatrix} 0 & -2.415 \\ 0 & 1 \end{bmatrix} \tag{3-70}$$

Then the product $[d][S]$ is formed:

$$[d][S] = [d]_1 = \begin{bmatrix} 1 & 1 \\ 1 & 3 \end{bmatrix}\begin{bmatrix} 0 & -2.415 \\ 0 & 1 \end{bmatrix} = \begin{bmatrix} 0 & -1.415 \\ 0 & 0.585 \end{bmatrix}$$

Now we may perform the iteration process on the new dynamical matrix $[d]_1$, which has the fundamental mode and frequency removed. Assuming, as before, a column matrix

$$\{x\}_0 = \{1 \quad 1\}$$

we obtain

$$[d]_1\begin{bmatrix} 1 \\ 1 \end{bmatrix} = \begin{bmatrix} -1.415 \\ 0.585 \end{bmatrix} = 0.585\begin{bmatrix} -2.42 \\ 1 \end{bmatrix}$$

No further iterations are necessary. Hence,

$$\frac{2g}{l\omega_2^2} = 0.585$$

or

$$\omega_2 = 1.85\sqrt{g/l} \tag{3-71}$$

which is the exact value.

The modal columns $\{A^{(r)}\}$ used in the preceding analysis are undetermined to the extent that we know only the ratios of the various amplitudes. Therefore, if we introduce a new arbitrary constant, C_r, for each mode, which may be evaluated for the initial conditions of the problem, the complete solution (Eq. 3–46) may be written in general as

$$\{q\} = \sum_{r=1}^{n} C_r\{A^{(r)}\} \sin(\omega_r t + \alpha_r) \tag{3–72}$$

This may, in turn, be written in complete matrix form as

$$\{q\} = [A][\sin(\omega t + \alpha)]\{C\} \tag{3–73}$$

Multiplying by the inverse of the matrix $[A]$, we have

$$\{y\} = [A]^{-1}\{q\} = [\sin(\omega t + \alpha)]\{C\} \tag{3–74}$$

Written out, these equations are

$$\begin{aligned} y_1 &= C_1 \sin(\omega_1 t + \alpha_1) \\ y_2 &= C_2 \sin(\omega_2 t + \alpha_2) \\ &\cdots\cdots\cdots \\ y_n &= C_n \sin(\omega_n t + \alpha_n) \end{aligned} \tag{3–75}$$

These equations are all independent; therefore, the quantities y_n are the normal coordinates of the system (see Art. 2–2).

Many more applications of matrix algebra to dynamics problems in aeronautics may be found in References 4–6 as well as in the subsequent chapters of this book.

PROBLEMS

3–1. Evaluate the fourth-order determinant

$$\begin{vmatrix} 0 & 1 & 3 & 5 \\ 2 & -4 & 6 & 3 \\ 1 & 3 & 0 & -4 \\ -2 & 5 & 3 & 1 \end{vmatrix}$$

3–2. Perform the following matrix operations: $[a] + [b]$, $[a] - [b]$, $[a][b]$, $[b][a]$, where $[a]$ and $[b]$ are given by

$$[a] = \begin{bmatrix} 1 & 2 \\ 3 & -4 \end{bmatrix}, \quad [b] = \begin{bmatrix} 2 & 3 \\ 5 & 0 \end{bmatrix}$$

3–3. For the matrix

$$[c] = \begin{bmatrix} -2 & 3 & 4 \\ 1 & 0 & 2 \\ 3 & -1 & 5 \end{bmatrix}$$

find $[c]'$ and $[c]^{-1}$.

3–4. For the two-degree-of-freedom system shown in Fig. 2–2, determine the inertia matrix, the stiffness matrix, the dynamical matrix, and the flexibility matrix.

3–5. For the system of Problem 4, assume that

$$m_1 = m_2 = m, \text{ and } k_1 = k_2 = k_3 = k.$$

Using matrix techniques, show that the frequency of the fundamental mode is given by $\omega_1^2 = k/m$.

3–6. Find the three modes and frequencies of a triple pendulum similar to the numerical example of Art. 3–6.

3–7. Using matrix techniques, find the two principal frequencies for the system described in Problem 7 of Chapter 2.

REFERENCES

1. Frazer, R. A., Duncan, W. J., and Collar, A. R. *Elementary Matrices.* New York: Cambridge University Press, 1950.
2. Aitken, A. C. *Determinants and Matrices.* New York: Interscience Publishers, Inc., 1949.
3. Pipes, L. A. *Applied Mathematics for Engineers and Physicists.* New York: McGraw-Hill Book Co., Inc., 1946. Ch. IV.
4. Benscoter, S. U., and Gossard, M. L. "Matrix Methods for Calculating Cantilever–Beam Deflections," *NACA Tech. Note 1827* (March, 1949).
5. Scanlan, R. H., and Rosenbaum, R. *Introduction to the Study of Aircraft Vibration and Flutter.* New York: The Macmillan Co., 1951. *(Dover reprint)*
6. Bisplinghoff, R. L., Ashley, H., and Halfman, R. L. *Aeroelasticity.* Reading, Mass.: Addison-Wesley Publishing Co., Inc., 1955.

CHAPTER 4

SELF-EXCITED VIBRATIONS AND STABILITY OF MOTION

4–1. Self-excited Vibrations. Chapters 1 and 2 dealt with the vibrations of both lumped and distributed mass systems—both free and forced vibrations were considered. The forced vibrations resulted from the application of an alternating force which was independent of the coordinates of the system but which depended only on the time. Thus the external force existed completely independently of the motion.

Quite often in mechanical (and electrical) systems, alternating forces exist which are *not* independent of the coordinates of the system. The vibrations resulting from such a situation are called *self-excited vibrations.* It is possible for the disturbing force to be an external one which depends on the coordinates of the system, or the disturbing force may actually be created by the motion itself; when the motion ceases, the exciting force no longer exists.

For example, suppose that we have a single-degree-of-freedom system with a *negative* viscous damping:

$$m\ddot{x} - c\dot{x} + kx = 0, \qquad c > 0 \tag{4–1}$$

The solution of this differential equation is (compare with Eq. 1–16b)

$$x = e^{+nt}(A \cos \bar{\omega}t + B \sin \bar{\omega}t) \tag{4–1a}$$

Because the sign of the exponential is now positive, the amplitude of vibration will increase with time.

More generally, we may consider a simple, linear, damped system acted upon by an external force which is a function of the coordinates:

$$m\ddot{x} + c\dot{x} + kx = F(x, \dot{x}, \ddot{x}) \tag{4–2}$$

Such a system will undergo self-excited vibration if the forcing function is of such a form that energy is fed into the motion. The frequency of self-excited vibration is clearly the natural frequency of the system.

Perhaps it is worthwhile to comment additionally on the energy transformation involved here, inasmuch as a similar discussion will take place in connection with flutter in Chapter 5. When the damping is positive, in a simple system such as we are discussing, the damping force does negative work on the system; this means that energy is removed from the system, usually in the form of heat. Thus, each successive cycle of the motion is reduced in amplitude and therefore possesses less kinetic energy; the decrease in kinetic energy is accounted for by the damping force. On the other hand, when the damping is negative, the damping force does positive work on the system; that is, each successive cycle of the motion is increased in amplitude and therefore possesses more kinetic energy. The increase in kinetic energy is provided by the negative damping force which must transmit this energy from an external source; without this external source of energy the motion cannot exist as a self-excited one.

One may consider the vibrations of systems with damping from the viewpoint of *dynamic stability*. For a case in which the amplitudes diminish with time (positive damping), we have a dynamically *stable* system; for a case in which the amplitudes increase with time (negative damping), we have a dynamically *unstable* system. Thus, in a complex vibratory system, where there is a possibility of energy being fed into the motion, we must determine the stability of the motion; a mathematical criterion for stability will be developed in Art. 4–2. Some additional

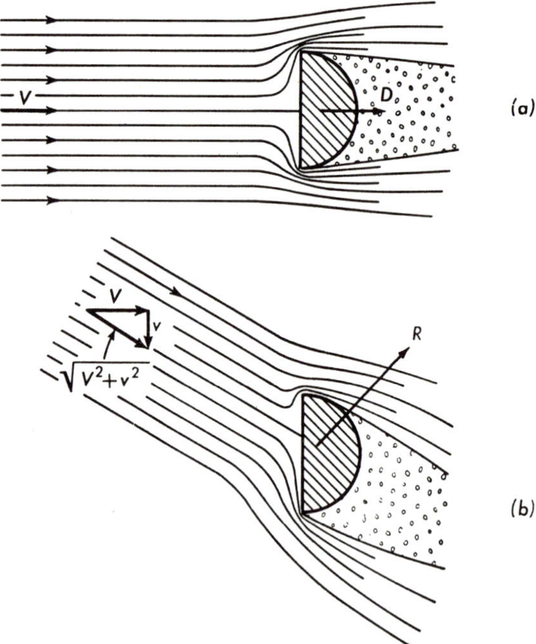

FIG. 4–1. "Half-round" body in a uniform airstream: (a) rest; (b) up-stroke of vertical motion.

examples here may help to illustrate vibrations of self-excited systems and emphasize the importance of the stability concept.

Bodies placed in a uniform airstream may exhibit self-excited vibrations under certain conditions. Consider, for instance, a "half-round" body placed in a uniform stream as shown in Fig. 4–1a. The only aerodynamic force acting on the body is a drag force. Suppose, however, that the body has been set into a vertical motion (normal to the airstream) by a gust or by some other means; Fig. 4–1b shows the flow pattern on the up-stroke.

It is assumed that the body is elastically restrained, however, so that a vibration will ensue, the amplitude of which is governed by the condition that the elastic restoring force becomes equal to the exciting force at some point in the motion. The spacing of the streamlines indicates a lowered pressure on the top of the body so that the resultant force on the body will be upward and back. On the downstroke, the situation will be exactly the reverse. Therefore, the vertical component of the resultant force will always tend to "reinforce" the motion (i.e., to feed energy into the motion). The formation of ice on a long wire results in a dynamically unstable cross-section which behaves similarly to that discussed above; therefore, the *galloping transmission line* phenomenon is one of self-excited vibration (Reference 1).

Let us put this analysis on a somewhat more mathematical basis. The resultant force R may be considered as being composed of vectors L and D normal and parallel to the relative wind, respectively, as shown in Fig. 4–2. The vertical force, R_v, may be expressed as

$$R_v = L \cos \alpha - D \sin \alpha \tag{4–3}$$

If R_v is upward when the motion is upward, and downward when the motion is downward, the situation will be one of self-excited vibration. We note, however, that the resultant force, and hence L and D, depends upon the angle of the relative wind α, and that α itself is variable. Thus

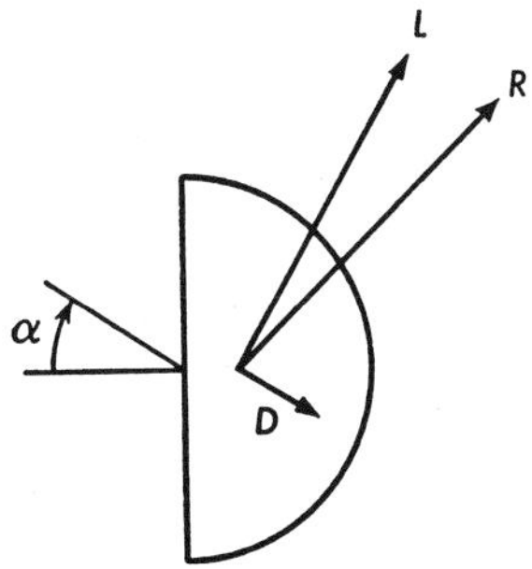

FIG. 4–2. Force diagram for "half-round" body.

we may formulate a stability criterion based upon the variation of the upward force R_v with the angle α. Therefore, if α is *negative*, as defined in Fig. 4–2, we see that for stable motion a negative change in R_v must be associated with a positive change in α. Mathematically,

$$\begin{aligned} \frac{dR_v}{d\alpha} &< 0 \Rightarrow \text{unstable motion} \\ \frac{dR_v}{d\alpha} &> 0 \Rightarrow \text{stable motion} \end{aligned} \tag{4–4}$$

From Eq. 4–3 we have, recalling that α is negative as shown in Fig. 4–2,

$$\frac{dR_v}{d\alpha} = \frac{dL}{d\alpha} \cos \alpha - L \sin \alpha + \frac{dD}{d\alpha} \sin \alpha + D \cos \alpha \tag{4–5}$$

For small values of α this may be written as

$$\frac{dR_v}{d\alpha} = \frac{dL}{d\alpha} + D \tag{4–6}$$

Thus the motion will be stable only for

$$\frac{dL}{d\alpha} + D > 0 \tag{4–7}$$

so that in order to ensure stability, the negative slope of the lift curve must be less than the ordinate of the drag curve, at any given value of the angle α.

A study of lift and drag curves for a variety of sectional shapes will quickly reveal that usually the lift curve rises from small values to some maximum with increasing α and then decreases. The drag, on the other hand, is always positive. Therefore, it is entirely likely that during some parts of the motion, the aerodynamic force will do positive work, while during other parts of the motion it will do negative work. This is shown typically in Fig. 4–3. A final discussion of this motion could now be made on the basis of the net work done per cycle (that is, whether positive or negative), but it is perhaps better to defer this to Chapter 5 where the flutter of airplane wings is discussed in detail.

In the preceding example, the external forces (aerodynamic in origin) depend upon the coordinates of the system and lead to a self-excited vibration. The *flutter* of aircraft wings and tails is not unlike the vibration phenomenon just described. The spectacular failure of the Tacoma Narrows Bridge in 1940 was due in large part to large-amplitude, self-excited vibrations of aerodynamic origin.

There is, however, another type of aerodynamic excitation different from that just discussed. When a fluid flows past a cylindrical object, it has been observed that the wake is not smooth but is composed of rather large eddies or vortexes which are shed from opposite sides of the body in a regular but alternating manner.* These vortexes are alternately clockwise and counterclockwise, which results in a change in aerodynamic circulation about the body and consequently produces an alternating force applied to the body in a direction normal to the free-wind direction, in accordance with the Kutta-Joukowski theorem of lift. This phenomenon is known as the Kármán vortex street and gives rise to high-frequency vibrations of the body. As an example, the singing of telephone wires is due to this phenomenon.

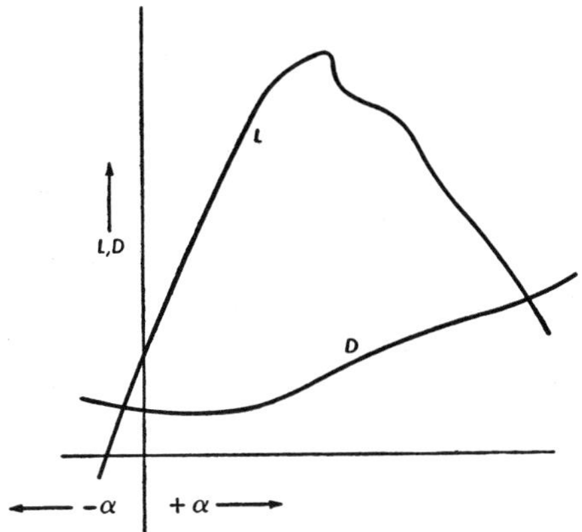

FIG. 4–3. Typical lift and drag curves.

It is interesting to compare the growth of amplitude with time for resonant and for self-excited vibrations. For vibrations at resonance, the amplitude increases according to κt (see Fig. 1–8), while for self-excited vibrations (characterized by Eq. 4–2) the amplitude increases according to κe^{nt}. Even if n were a very small quantity, after a given length of time the amplitude resulting from the self-excited vibration would be many times as large as that resulting from vibration at resonance. Flutter of aircraft components is characterized by motion according to κe^{nt}; this rapid build-up of amplitude accounts for what has been termed the "explosive nature of flutter."

Other cases of self-excited vibration have been found in engines and governors, many types of electrical devices (vacuum tubes, especially), various types of rotating-shaft vibration, mechanical devices of all kinds, several hydraulic devices, and almost all musical instruments. The squeaking door hinge is an everyday example, as is the piece of chalk

* Wakes in real fluids are, of course, never "smooth" in the strict sense, but in many instances the eddy size is so small that at least the wake appears somewhat uniform.

held perpendicular to the blackboard. The latter two examples arise from the variable friction characteristics which they possess.

4–2. Stability of a Given State of Motion. Let us consider for a moment the meaning of the word *stability* in a general sense. If a system is in a certain state of equilibrium, any disturbance of finite magnitude applied to the system will cause a free motion of the system following the disturbance; if this free motion ultimately disappears, the motion is said to be stable. The equilibrium may be such that the disturbance creates forces within the system that *tend* to return the system to its original configuration—such a system is said to be statically stable. For example, the simple pendulum possesses two equilibrium configurations, one when it is in its normal downward position and the other when it is inverted; the first is a statically stable equilibrium configuration, while the second is a statically unstable equilibrium configuration. A ball rolling on a horizontal frictionless table is in a statically neutral equilibrium configuration.

Once the system is in a state of statically stable equilibrium, it is of importance to observe the motion following a disturbance of the equilibrium. If this resulting motion is such that the system returns *to* its equilibrium position, the system is dynamically stable; if the system does not return to its equilibrium position, the system is dynamically unstable.

Thus, in the case of an airplane, we may consider some steady, known flight path as representing an equilibrium state. A disturbance applied to the airplane, such as a gust or an abrupt movement of the controls, will cause a deviation from the equilibrium state, and the subsequent free motion will be characterized by the stability characteristics of the airplane. We observe at once that static stability is requisite for dynamic stability, although the mere fact that a system is statically stable does not in itself insure dynamic stability.

The stability characteristics of a given system may, of course, be determined from a study of the solutions of the governing differential equations of motion; however, it is also possible to determine the stability of a system directly from the frequency equation.*

We deal with the small vibrations of linear systems so that the differential equations of motion take the form

$$\begin{aligned} m_1\ddot{x}_1 + c_1\dot{x}_1 + k_1x_1 + k_1'x_2 + \cdots &= 0 \\ &\vdots \\ m_n\ddot{x}_1 + c_n\dot{x}_1 + k_nx_1 + k_n'x_2 + \cdots &= 0 \end{aligned} \tag{4–8}$$

* E. J. Routh, "On the Stability of a Given State of Motion," *Adams Prize Essay*, 1877; also, E. J. Routh, *Dynamics of a System of Rigid Bodies*, (New York: The Macmillan Co., 6th ed., 1892), p. 221.

As before, we assume solutions of the form

$$x_1 = A_1 e^{pt} \\ \cdot \\ \cdot \\ x_n = A_n e^{pt} \tag{4–9}$$

Introduction of these assumed solutions into the equations of motion (Eq. 4–8) will result in an nth degree equation for the square of the frequency p. The roots $p_1 \cdots p_n$ of the frequency equation will, in general, be complex quantities of the form

$$p = q + ir \tag{4–10}$$

Therefore, if we write the solutions (Eq. 4–9) in the form

$$x_1 = C_1 e^{p_1 t} + C_2 e^{p_2 t} + \cdots + C_n e^{p_n t}$$

or (4–11)

$$x_1 = e^{q_1 t}(A_1 \sin r_1 t + B \cos r_1 t) + \cdots$$

we see that the real part of the complex quantity (Eq. 4–10) determines the damping, and the imaginary part determines the frequency of vibration. *For the motion to be stable, the real parts of all the values of* p *must be negative.*

Therefore, if rules for deciding stability are to be determined, we must be concerned with the roots of the frequency equation. For example, consider the cubic frequency equation

$$p^3 + a_2 p^2 + a_1 p + a_0 = 0 \tag{4–12}$$

A cubic equation will have, in general, one real root and two complex conjugate roots; therefore,

$$p_1 = q_1 \\ p_2 = q_2 + ir_2 \\ p_3 = q_2 - ir_2 \tag{4–13}$$

The frequency equation may be written in terms of the three roots p_1, p_2, p_3 as

$$(p - p_1)(p - p_2)(p - p_3) = 0 \tag{4–14}$$

which then gives

$$a_2 = -(q_1 + 2q_2) \\ a_1 = q_2^2 + 2q_1 q_2 + r_2^2 \\ a_0 = -q_1(q_2^2 + r_2^2) \tag{4–15}$$

The criterion for stability is that $q_1 < 0$, $q_2 < 0$ (from Eq. 4–13); therefore, from Eq. 4–15 we have that a_0, a_1, a_2 must all be positive.

The boundary between stability and instability (i.e., *neutral* stability) is given by the condition that the real parts of p be zero.

A study of Eq. 4–15 reveals that it is quite possible for a_0, a_1, a_2 all to be positive, and yet q_2 may not be negative. Thus the condition that the coefficients a_0, a_1, a_2 be positive is only a *necessary* condition if the motion is to be stable. Suppose that $q_2 = 0$ (neutral stability). Then

$$a_2 = -q_1, \quad a_1 = r_2^2, \quad a_0 = -q_1 r_2^2$$

Upon eliminating q_1 and r_2 we find that $a_0 = a_1 a_2$. Comparing this result with Eq. 4–15 shows that if q_2 is negative, a_0, a_1, a_2 are all positive and also

$$a_1 a_2 > a_0 \tag{4–16}$$

This then constitutes the *sufficient* condition for stability. The necessary and sufficient conditions together form the *complete criterion* for stability.

An analysis similar to that above for quartic and quadratic frequency equations would give the following conditions on the coefficients:

Quartic Equation

$$p^4 + a_3 p^3 + a_2 p^2 + a_1 p + a_0 = 0 \tag{4–17}$$

The conditions for stability are

$$a_0, a_1, a_2, a_3 > 0$$

and (4–18)

$$(a_1 a_2 a_3) > (a_1^2 + a_0 a_3^2)$$

Quadratic Equation

$$p^2 + a_1 p + a_0 = 0 \tag{4–19}$$

The conditions for stability are

$$a_0, a_1 > 0 \tag{4–20}$$

The foregoing results stem from the work of Routh. An alternative formulation of the conditions for stability may be stated in terms of determinants.* For the quartic equation

$$a_4 p^4 + a_3 p^3 + a_2 p^2 + a_1 p + a_0 = 0 \tag{4–21}$$

the four determinants

$$|a_1|, \quad \begin{vmatrix} a_1 & a_0 \\ a_3 & a_2 \end{vmatrix}, \quad \begin{vmatrix} a_1 & a_0 & 0 \\ a_3 & a_2 & a_1 \\ 0 & a_4 & a_3 \end{vmatrix}, \quad \begin{vmatrix} a_1 & a_0 & 0 & 0 \\ a_3 & a_2 & a_1 & a_0 \\ 0 & a_4 & a_3 & a_2 \\ 0 & 0 & 0 & a_4 \end{vmatrix} \tag{4–22}$$

* This result is known as the *Hurwitz criterion for stability.*

must all be positive. This stability condition is completely equivalent to the Routh conditions (Eq. 4–18).

The quartic equation appears so frequently in the study of aircraft dynamics that it is worthwhile to give a brief discussion of the information that may be obtained from it by inspection of the Routh discriminant $a_3a_2a_1 - a_1^2 - a_0a_3^2$. If all coefficients are positive and the discriminant is positive, there is no possibility of either a pure divergent motion or a divergent oscillatory motion. If the discriminant is zero, there will be an undamped oscillation, while if it is negative, there will be a divergent oscillatory motion. If the coefficient a_0 is zero, one of the roots is zero, and thus one of the modes can continue unchanged since one of the roots is zero. If any one of the coefficients is negative, there can be either a divergent oscillatory motion or a pure divergent motion in one of the modes.

PROBLEMS

4–1. A system is described by the equation of motion

$$mx + c\dot{x} + kx = F_0\dot{x}, \quad F_0 > 0$$

Solve this equation and discuss the stability of the motion for $F_0 > c$ and for $F_0 < c$ by means of the criteria developed in Art. 4–2. Sketch the displacement-time diagrams.

4–2. Show how an unstable motion may take place in a single-degree-of-freedom system possessing viscous damping, if the exciting force is proportional to the acceleration. The differential equation of motion is

$$m\ddot{x} + c\dot{x} + kx = P\ddot{x}$$

4–3. An airfoil is mounted in an airstream and restrained to move in the vertical direction only, as shown in Fig. 4–4. The aerodynamic lift and drag are given by

$$L = L_0 \sin 2\alpha, \quad D = D_0 \cos 2\alpha$$

The airfoil is observed to begin an unstable motion when $L_0/D_0 = \frac{1}{2}$ and θ is 15°. At what angle of attack α is the airfoil oriented? *Hint*: Neglect any restoring force provided by the mounting device.

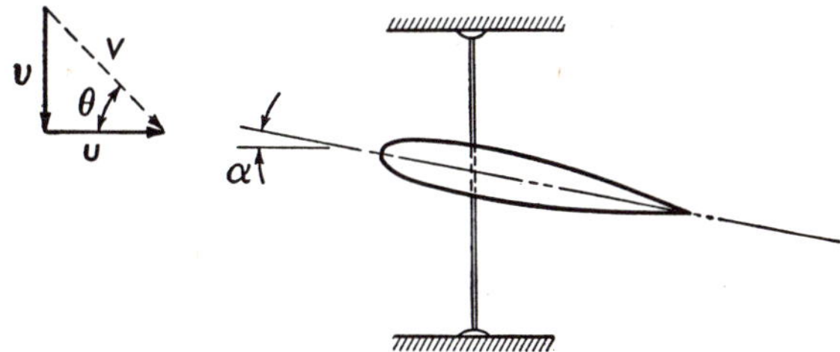

Fig. 4–4. Problem 3.

4–4. Fig. 4–5a shows the configuration of a "conventional" airplane landing-gear system (full-castering tail wheel), and Fig. 4–5b shows the configuration of a "tricycle" airplane landing-gear system (full-castering nose wheel). Consider in each case that some disturbance, such as a gust of air, has caused an angular displacement of the centerline of the airplane about the center of gravity. Explain why the motion, as a consequence of such a disturbance, is stable in one system and is unstable in the other system.

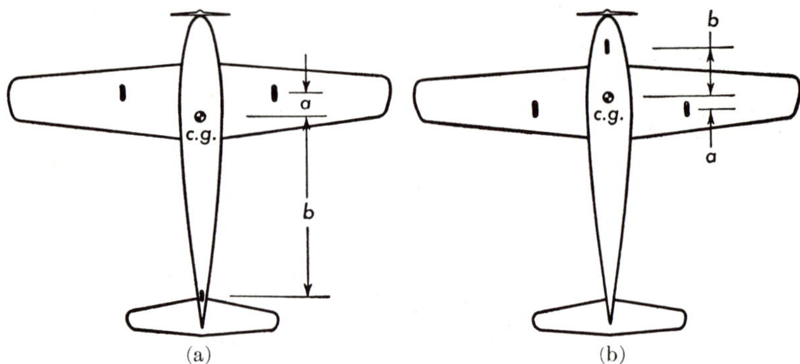

FIG. 4–5. Problem 4: (a) "conventional" landing-gear arrangement; (b) "tricycle" landing-gear arrangement.

4–5. The longitudinal flight stability of a certain airplane has been investigated, and the equation which describes the stability of motion was found to be

$$p^4 + 3.85p^3 + 5.02p^2 + p + 0.5 = 0$$

Is the motion stable or unstable?

4–6. Derive the stability conditions for a quadratic equation (Eq. 4–20) and for a quartic equation (Eq. 4–18).

REFERENCES

1. SCANLAN, R. H., and ROSENBAUM, R. *Introduction to the Study of Aircraft Vibration and Flutter*. New York: The Macmillan Co., 1951. *(Dover reprint)*
2. DEN HARTOG, J. P. *Mechanical Vibrations*. 4th ed. New York: McGraw-Hill Book Co., Inc., 1956.
3. WHITTAKER, E. T. *Analytical Dynamics of Particles and Rigid Bodies*. New York: Dover Publications, Inc., 1945.

CHAPTER 5

ELEMENTARY THEORY OF WING FLUTTER

5-1. Introduction. Recent years have witnessed continually increasing flight velocities of airplanes and missiles. These higher flight speeds have necessitated rapid developments in the science of aerodynamics because air can no longer be treated as an incompressible fluid in steady flow. The increased aerodynamic loads associated with the higher flight velocities, together with the thinner, more flexible wings required for flight at high speeds, soon indicated that the aeronautical designer could not continue to treat the various components of the aircraft as rigid structures, but instead must begin to take cognizance of their dynamical characteristics.

A number of interesting and important phenomena have thus been encountered which, because of the interaction between the aerodynamic aspects and the structural aspects which give rise to them, have been collected under the single name *aeroelastic effects*. Included within these phenomena are flutter, wing divergence, reduction of control effectiveness, etc. The first of these, flutter, will be studied briefly in the present chapter, while the remaining items are deferred until Chapter 6, where a brief and largely nonanalytical discussion of the general problems of static aeroelasticity will be given.

Flutter was observed many years ago,* but only recently has the problem risen to such extreme importance. The analysis of flutter is an extremely intricate and difficult procedure for a modern airplane or missile, especially if all aspects of the problem are considered. Such an analysis is beyond the scope of this volume; however, it is possible to treat the flutter problem in an elementary manner, placing emphasis on the formulation of the problem and the physical interpretation, while offering only general comments on the more detailed aspects of computational procedures and refinements to the basic theory.

5-2. The Mechanism of Flutter. Let us assume that an airplane wing in steady flight at some moderate airspeed is suddenly disturbed by, for example, a gust. The aerodynamic forces are often such as to tend to

* See References 1 and 2 for some historical background and for detailed lists of references to literature on flutter analysis.

increase the amplitude of vibration, while the elastic forces of the wing always tend to restore the wing to its original configuration; therefore, after some time the elastic forces will have caused the motion to disappear. Now suppose that the flight velocity is somewhat increased, whereupon the aerodynamic forces will be increased because they vary essentially with the square of flight velocity. There must be some velocity at which the aerodynamic forces (which excite the motion) will just balance the elastic forces (which attempt to resist the motion); a vibration of constant amplitude will result. The flight velocity at which this equilibrium of forces is attained is called the *critical flutter speed* of the wing. For velocities greater than the critical one, the exciting forces will exceed the restoring forces, and therefore the amplitude of the disturbance will grow without limit. Thus, we recognize that flutter is a self-excited vibration; i.e., the disturbing forces arise as a consequence of the motion itself.

Unfortunately, it is not possible to give a really complete picture of the mechanism of flutter based simply on the discussion of the unstable single-degree-of-freedom system treated in Chapter 4 (see Fig. 4–1) and in the preceding paragraph. Instead, we must look at the wing as a coupled two-degree-of-freedom system because bending vibrations will exist not independently but simultaneously with a twisting motion of the wing. Furthermore, it is possible that the aerodynamic forces will not always tend to excite the motion but will at times tend to damp it. The presence of an exciting aerodynamic force is a direct consequence of the dependent relationship between flexure and twist.

In order to illustrate this, assume that a wing is disturbed initially by being twisted, without flexure, to a larger angle of attack. The change in angle of attack will cause a change in lift in such a manner as to do work on the wing (positive work). If the twist is such as to cause a decrease in angle of attack, the "additional" lift will do work against the motion of the wing (negative work). Therefore, the torsional motion causes an exciting force during part of the cycle and a damping force during the remainder of the cycle. Now suppose that the wing is simply deflected upward (downward) in bending. The vertical translational velocity, added to the flight velocity, will give a resultant velocity vector directed so as to reduce (increase) the angle of attack; therefore, this motion gives rise to a damping force.

We see now that negative work is done on the wing by part of the torsional motion, by the flexural motion, and by the elastic restoring forces; positive work is done on the wing by part of the torsional motion. The motion will maintain itself (the condition for flutter) when the net positive work just balances the dissipation of energy due to all the damping forces. The magnitude of the positive work done by the additional lift due to the twist is directly dependent upon the phase relationship between the coupled torsional and flexural motions, as will be explained in the next paragraph.

Suppose that the twist lags *behind* the flexural displacement. Flexural displacement is defined as positive down, and torsional displacement is defined as positive when the angle of twist is a stalling angle. A motion in which twist lags flexure will be one for which a positive flexural displacement will have associated with it a negative angle of twist; a leading motion will be one for which a positive flexural displacement will have associated with it a positive angle of twist. Therefore, some time after the wing has begun an upstroke, a positive twist begins to grow; this twisting will aid in the upward motion. After the motion has reached its maximum amplitude and flexural displacement begins to decrease, the twist becomes negative, again aiding the motion. Therefore, it is clear that a motion wherein the twist lags flexure will produce positive work throughout the whole cycle. Similar reasoning for the case of leading phase will show that the net positive work is zero, or at best, is produced for only a small portion of the cycle.*

The pertinent facts concerning the mechanism of flutter may now be summarized as follows:

(a) The oscillation is self-excited.
(b) The lower boundary of the flutter region, called the *critical flutter speed*, is characterized by a vibration in which the exciting forces exactly equal the damping forces; such vibrations are of constant amplitude (determined by the magnitude of the initial disturbance).
(c) The motion must be both torsional and flexural in nature, and the torsional motion must lag behind the flexural motion.
(d) The flutter speed will depend in large measure upon the torsional and flexural stiffnesses of the wing.

Thus far we have discussed only bending-torsion flutter, but this is by no means the only type of flutter which may occur. Other types of binary flutter which may occur usually involve the rotation of some control surface about its own hinge line; for example, one might have wing bending-aileron flapping, fuselage torsion-rudder flapping, fin bending-rudder flapping, to mention only a few. Note that flutter is not restricted to wings but may occur for tails and other major components, as well.

It is also quite possible, and actually the much more commonly encountered case, for more than two modes of motion to be present.

* The concept of the magnitude of work being dependent on a phase relationship may already be familiar to the reader through study of the theory of alternating electrical circuits. There it is shown that power = voltage $\times$ current $\times \cos\theta$, where θ is the phase angle between the voltage and the current.

For example, wing torsion-wing bending-aileron flapping is a very common type of ternary flutter. Higher degree motions often exist which require that one consider not only the fundamental flexural and torsional modes of the wing but some of the higher modes as well, and perhaps also some of the fuselage vibration modes. The question of how many modes of vibration must be considered in a given flutter analysis depends upon the relative values of the vibration frequencies of the various modes and on the degree of refinement desired in the analysis. It is frequently found necessary to include also the rigid-body motions of the airplane as a whole.

There are yet other types of flutter which may depend on such flow phenomena as separation, turbulence, shock-wave action, and vortex shedding. These types of flutter—usually called "nonclassical" as contrasted with the "classical" type of flutter discussed in the preceding paragraphs—are not yet well understood. All further remarks in this chapter concerning flutter will refer to the classical type of flutter.

5–3. Mathematical Formulation of the Problem. The first task is to write the equations of motion of the system. The problem is twofold: First, we must determine the contributions of the elastic structure to the restoring forces, and second, we must determine the nature of the aerodynamic forces that are acting. The latter is not easy because the vibrating wing presents a problem in the aerodynamics of unsteady flow. It is apparent that the aerodynamic forces will depend upon the deflections, velocities, and accelerations of the structure as characteristic of a self-excited system. The velocities will give rise to the usual type of aerodynamic forces, while the accelerations will give rise to forces resulting from the inertia of the air set into motion by the vibrating wing. The aerodynamic forces will be considered in somewhat more detail in Art. 5–4.

In order to approach the problem in a reasonable manner, certain simplifying assumptions will be made.

(a) The wing is assumed to be of infinite aspect ratio so that the aerodynamic problem is not complicated by the effects of finite span.
(b) Small amplitudes of vibration are assumed throughout.
(c) Potential (ideal) incompressible fluid flow is assumed.
(d) A representative portion of the wing of unit span will be considered.*
(e) Only bending-torsion flutter will be considered.

The analysis which follows is a much simplified version of that presented by Theodorsen (see References 3 and 4).

* This is often taken as the ¾-span station.

The wing section of unit span is shown in Fig. 5–1 in its deflected position. The vertical displacement is defined by h (positive down) and the angular displacement by α (stalling angle positive). The semichord is denoted by b. The distance a measures the location of the point c.t. (in percentage of b) with respect to the mid-chord (positive aft of the mid-chord point). The point c.t., called the *center of twist*,* designates that point on the section where the application of a force will cause a translation but no rotation, and conversely, where the application of a torque will cause a rotation but no translation.

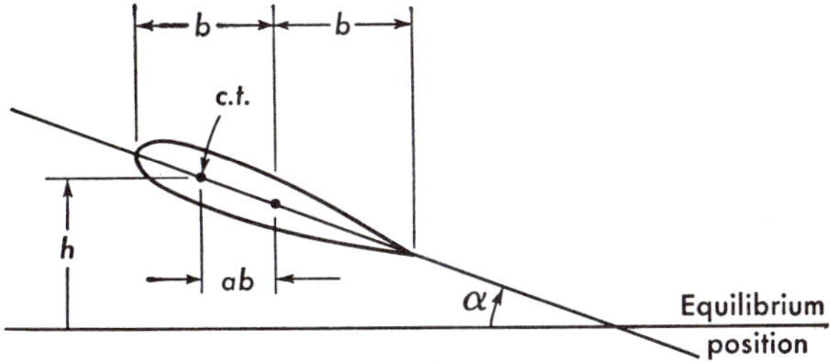

FIG. 5–1. Wing section displaced from equilibrium.

The kinetic energy of an element of mass dm located at a distance r from the point c.t. may be written as

$$dT = \tfrac{1}{2}(\dot{h} + r\dot{\alpha})^2 \, dm$$

The total kinetic energy may be obtained simply by integration from leading edge to trailing edge. Thus,

$$T = \tfrac{1}{2}M_w\dot{h}^2 + S_w\dot{h}\dot{\alpha} + \tfrac{1}{2}I_w\dot{\alpha}^2 \tag{5–1}$$

where

$$M_w = \int_0^{2b} dm = \frac{\text{wing mass}}{\text{unit span}}$$

$$S_w = \int_0^{2b} r \, dm = \frac{\text{wing statical moment}}{\text{unit span}} \tag{5–2}$$

$$I_w = \int_0^{2b} r^2 \, dm = \frac{\text{wing mass moment of inertia}}{\text{unit span}}$$

The bending and torsion stiffnesses of the structure may be considered in terms of their "equivalent spring constants," which may be denoted by C_h for bending and C_α for torsion. The potential energy is therefore

$$U = \tfrac{1}{2}C_h h^2 + \tfrac{1}{2}C_\alpha \alpha^2 \tag{5–3}$$

* Sometimes called *shear center* or *flexural center*. (For the simplified analysis presented here, these terms may be used interchangeably.)

The kinetic and potential energies may be introduced into Lagrange's equation (Eq. 2–8) with h and α as the generalized coordinates, yielding the two differential equations of motion*

$$M_w\ddot{h} + S_w\ddot{\alpha} + C_h h = 0$$
$$I_w\ddot{\alpha} + S_w\ddot{h} + C_\alpha \alpha = 0 \tag{5–4}$$

These are the differential equations which would hold for a wing vibrating in a vacuum (i.e., no aerodynamic forces) with no damping of any kind present.† Even if there are no aerodynamic forces acting, however, the above equations do not apply exactly because there will be present some damping which arises from the structure itself. This damping comes not only from the frictional forces at joints, but also from the internal friction of the material of which the structure is composed.

This *structural damping* is very difficult to evaluate by analytical means; however, the results of numerous experiments have revealed considerable information concerning the nature of this damping (see References 5 and 6). In particular,

(a) Structural damping is a function of the amplitude of vibration of the elastic system.
(b) Structural damping is independent of the frequency of vibration.
(c) The structural damping force is in phase with the velocity of motion.

The fact that the structural damping is proportional to amplitude but in phase with velocity requires that special consideration be given to the formulation of those terms in the differential equations of motion which depend on this damping.

If the dimensionless coefficient of structural damping is denoted by g, we may write the magnitudes of the damping forces as $g_h C_h h$ and $g_\alpha C_\alpha \alpha$. In order to make these forces in phase with the velocities we introduce the assumption of simple harmonic motion:‡

$$h = h_0 e^{i\omega t}, \quad \alpha = \alpha_0 e^{i(\omega t+\theta)} \tag{5–5}$$

where ω is the frequency of vibration, θ is the phase angle by which flexure leads twist, and h_0 and α_0 are the amplitudes of vibration. Therefore, in order for the damping forces to be in phase with velocity, we must multiply by $i = \sqrt{-1}$, so that the damping forces become $ig_h C_h h_0$ and $ig_\alpha C_\alpha \alpha_0$.

* Note that these two equations are coupled through their inertia terms rather than through their elastic terms. This situation was also encountered in Problem 5 of Chapter 2. Compare also with Eq. 3–9.

† Compare with Problem 8, Chapter 2.

‡ Note that this way of formulating simple harmonic motion is the same as that used in previous chapters, since $e^{i\omega t} = \cos \omega t + i \sin \omega t$.

Damping forces of this form may be put into the equations of motion (Eq. 5–4) by means of a *dissipation function* (see page 34) defined by

$$D = \frac{1}{2}\frac{g_h C_h \dot{h}^2}{\omega} + \frac{1}{2}\frac{g_\alpha C_\alpha \dot{\alpha}^2}{\omega} \tag{5-6}$$

The differential equations of motion then become

$$\begin{aligned} M_w\ddot{h} + S_w\ddot{\alpha} + \frac{1}{\omega} g_h C_h \dot{h} + C_h h &= 0 \\ I_w\ddot{\alpha} + S_w\ddot{h} + \frac{1}{\omega} g_\alpha C_\alpha \dot{\alpha} + C_\alpha \alpha &= 0 \end{aligned} \tag{5-7}$$

The lift force and moment acting on the wing section (per unit span) are denoted by P and M. The complete form of the equations of motion is then

$$\begin{aligned} M_w\ddot{h} + S_w\ddot{\alpha} + g_h \frac{C_h}{\omega}\dot{h} + C_h h &= P \\ I_w\ddot{\alpha} + S_w\ddot{h} + g_\alpha \frac{C_\alpha}{\omega}\dot{\alpha} + C_\alpha \alpha &= M \end{aligned} \tag{5-8}$$

It will be convenient to introduce the undamped natural frequencies for the case of *uncoupled* vibrations. For bending and torsional vibrations which exist independently of each other, the uncoupled equations of motion are

$$\begin{aligned} M_w\ddot{h} + C_h h &= 0 \\ I_w\ddot{\alpha} + C_\alpha \alpha &= 0 \end{aligned}$$

The natural frequencies (undamped independent vibrations) are then

$$\omega_h^2 = \frac{C_h}{M_w}, \quad \omega_\alpha^2 = \frac{C_\alpha}{I_w} \tag{5-9}$$

The differential equations of motion for the system may now be written in final form as

$$\begin{aligned} \ddot{h} + \frac{S_w}{M_w}\ddot{\alpha} + g_h \frac{\omega_h^2}{\omega}\dot{h} + \omega_h^2 h &= \frac{P}{M_w} \\ \ddot{\alpha} + \frac{S_w}{I_w}\ddot{h} + g_\alpha \frac{\omega_\alpha^2}{\omega}\dot{\alpha} + \omega_\alpha^2 \alpha &= \frac{M}{I_w} \end{aligned} \tag{5-10}$$

In general, the aerodynamic terms P and M will be functions of h and α and their time derivatives; therefore, Eq. 5–10 may be made homogeneous by combining terms and can be solved by introducing the assumption of simple harmonic motions (Eq. 5–5). The expressions for P and M will be given in Art. 5–4.

The introduction of the simple harmonic expressions (Eq. 5–5) thus yields two *algebraic* equations for the unknown amplitudes h_0 and α_0. These algebraic equations will be of the form

$$\begin{aligned} Ah_0 + B\alpha_0 &= 0 \\ Ch_0 + D\alpha_0 &= 0 \end{aligned} \tag{5–11}$$

where the coefficients A, B, C, and D are, in general, complex. In order for solutions of these equations to exist, other than $h_0 = \alpha_0 = 0$, the determinant of the coefficients must vanish (see Art. 2–2). Thus,

$$\begin{vmatrix} A & B \\ C & D \end{vmatrix} = 0 \tag{5–12}$$

This determinant yields a *frequency equation* for the frequency of vibration ω. Unfortunately, the quantities $A \cdots D$ are complex quantities containing not only contributions from the elastic, inertia, and damping terms, but also from the aerodynamic terms. It will be shown in the next article that the aerodynamic terms are functions of the flutter velocity V as well as the flutter frequency ω; therefore Eq. 5–12 has both velocity and frequency as its unknowns. Inasmuch as the resulting equation is composed of complex quantities, we may utilize the statement that for a complex quantity to be zero, both the real and imaginary parts must be zero. Thus Eq. 5–12 is in reality two equations, each of which is equal to zero.

The solution of the determinant (Eq. 5–12), called the *flutter determinant* (or stability determinant), is the focal point of the flutter analysis. The evaluation of this determinant will be discussed briefly in Art. 5–5.

5–4. Aerodynamic Force and Moment. We now have before us the task of determining expressions for the aerodynamic force and moment, P and M.

A wing in a steady incompressible ideal flow at constant angle of attack has a lift force acting on it given by the Kutta-Joukowski theorem as

$$P = \rho V \gamma \tag{5–13}$$

where P is the lift force per unit span, ρ is the fluid density, V is the free stream velocity, and γ is the circulation.

At the beginning of motion, a *starting vortex* is shed from the trailing edge of the wing which has a circulation of $-\gamma$. Thereafter, each time that the angle of attack of the wing is changed, a new value of the circulation is created about the wing, due to the shedding of an additional vortex from the trailing edge. Thus, for an oscillating airfoil, the angle of attack will be changing continually, thereby causing a continuous "sheet" of vortexes to be shed. This will cause a continuous change in circulation, and hence the lift force on the airfoil will be a function of time.

Although this simple explanation does show the time-dependent nature of the aerodynamic problem of the oscillating wing, the picture is not complete. First, the continuous vortex sheet which is shed from the trailing edge of the wing induces velocities in the neighborhood of the wing which vary with time and the frequency of vibration. Thus, the effect of the wing *wake* is to be considered. Second, a body with zero circulation around it may experience forces and moments if the body is being accelerated. These forces and moments, arising from the inertia of the fluid which is set into motion by the movement of the body, are called *aerodynamic inertia forces* or sometimes *apparent mass forces.**

It should now be clear that the analysis of the aerodynamics of an oscillating airfoil is not an easy problem. Theodorsen (Reference 3) was able, however, to obtain expressions for the lift and moment of an oscillating airfoil after making certain simplifying assumptions.

In Art. 5–2 we found that for certain conditions the airfoil would perform a rather orthodox damped motion because the restoring forces were able to overcome the exciting forces. For other conditions, the airfoil would undergo self-excited vibrations of rapidly increasing amplitude. When the conditions were such that the restoring forces exactly equaled the exciting forces, the motion was sustained at a certain constant amplitude. The frequency of vibration at this condition is called the *flutter frequency*. Theodorsen assumed that at this boundary condition (that is, between stable and unstable motions) the motion would be simple harmonic in nature. The velocity at which this condition occurs is called the *flutter velocity*.

Considering the fluid to be ideal and the airfoil to be undergoing simple harmonic oscillations, Theodorsen set up velocity potentials to account for torsional displacement and velocity (φ_α and $\varphi_{\dot\alpha}$), for flexural velocity ($\varphi_{\dot h}$), and for the effects of the wake (φ_γ), i.e., for the circulation effects. These velocity potentials must then satisfy Laplace's equation

$$\frac{\partial^2\varphi}{\partial x^2} + \frac{\partial^2\varphi}{\partial y^2} = 0$$

which governs the unsteady flow of an ideal fluid. This potential equation is a linear one, and therefore the various potentials may be added together to give the complete solution. The several potentials are related to each other and to the airfoil problem by the *Kutta condition*, which states that the flow velocity at the sharp trailing edge is finite. Thus, there is

* See, e.g., W. F. Durand, *Aerodynamic Theory*, Vol. I (Berlin: Springer Publishing Co., 1934), I, pp. 242–245.

the additional equation

$$\frac{\partial}{\partial x}(\varphi_\alpha + \varphi_{\dot\alpha} + \varphi_{\dot h} + \varphi_\gamma) = \text{finite value}$$

which the potentials must satisfy at the trailing edge.

With known expressions for the velocity potentials, it is possible to determine the velocity distribution around the airfoil and to determine the pressure distribution by application of Bernoulli's equation. Integration of the pressures over the surface of the airfoil then yields the force and moment as a function of the frequency of vibration of the wing and of the time.

Therefore, we see that the aerodynamic analysis is quite complex; it must account not only for the pressure distributions of steady flow, but also for those due to the wake and the apparent mass effects.

The Theodorsen analysis just described is much too lengthy and involved to present here, and therefore only the final results will be set down:

$$P = -\pi\rho b V^2 \left\{ \frac{b}{V^2}\ddot{h} + \frac{2}{V}C(k)\dot{h} - \frac{b^2 a}{V^2}\ddot{\alpha} + [1 + (1 - 2a)C(k)]\frac{b}{V}\dot{\alpha} + 2C(k)\alpha \right\} \tag{5-14}$$

$$M = -\pi\rho b^2 V^2 \left\{ -\frac{ab}{V^2}\ddot{h} - (1 + 2a)\frac{1}{V}C(k)\dot{h} + \left(\frac{1}{8} + a^2\right)\frac{b^2}{V^2}\ddot{\alpha} - \left[a - \frac{1}{2} + 2\left(\frac{1}{4} - a^2\right)C(k)\right]\frac{b}{V}\dot{\alpha} - (1 + 2a)C(k)\alpha \right\} \tag{5-15}$$

These expressions, cumbersome as they are, are even more complicated by the function $C(k)$, where k is defined by

$$k = \frac{b\omega}{V} \tag{5-16}$$

and is called the *reduced frequency*. The function $C(k)$, which arises from the circulation terms in the aerodynamic analysis, is an extremely complicated one, and furthermore it is a complex quantity. Separating real and imaginary parts, $C(k)$ may be written as

$$C(k) = F(k) + iG(k) \tag{5-17}$$

$F(k)$ and $G(k)$, which are made up of combinations of Bessel functions, may be represented as shown in Fig. 5–2; numerical values are given in Table 5–1. An extensive set of tables for the Theodorsen function are given in Reference 7.

The preceding equations for aerodynamic force and moment show fairly clearly the dependence of those quantities on the displacement

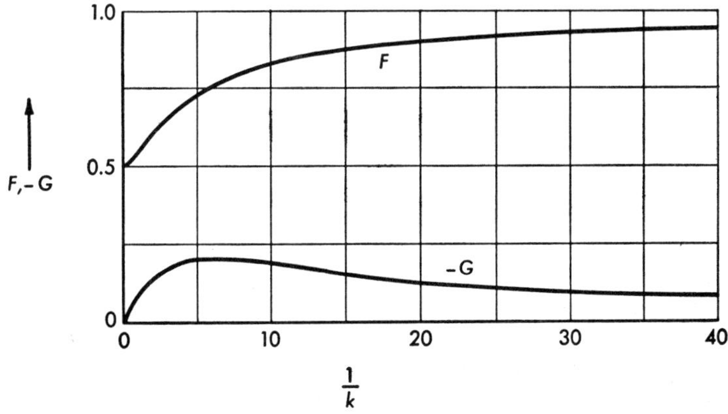

FIG. 5-2. Theodorsen $F(k)$ and $G(k)$ functions for harmonic motion.

coordinates and their time derivatives. Even without a derivation of these expressions one sees clearly the terms describing the apparent mass forces, the forces caused by the displacement velocities, and the static forces arising from a change in angle of attack.

With the expressions (Eqs. 5-14 and 5-15) for P and M, it is now clear that introduction of the simple harmonic expressions (Eq. 5-5) will lead to algebraic equations of the form (Eq. 5-11) and thence to the flutter determinant. The solution of the flutter determinant then will yield the

TABLE 5-1

VALUES OF THE THEODORSEN FUNCTIONS $F(k)$ AND $G(k)$

k	F	$-G$
0	1.000	0
0.05	0.909	0.130
0.10	0.832	0.172
0.20	0.728	0.189
0.30	0.665	0.179
0.40	0.625	0.165
0.50	0.598	0.151
0.60	0.579	0.138
0.80	0.554	0.116
1.00	0.539	0.100
1.20	0.530	0.088
1.50	0.521	0.0736
2.00	0.513	0.0577
3.00	0.506	0.0400
4.00	0.504	0.0305
6.00	0.502	0.0206
10.00	0.501	0.0124
∞	0.500	0

flutter frequency ω, and for a particular value of k, the flutter velocity V may be obtained.

It should be noted that the final form of the equations of motion, obtained from combining Eq. 5–10 with Eqs. 5–14 and 5–15, will be coupled elastically as well as inertially (see footnote on page 35). Coupling will also arise through the damping terms. The additional coupling between the equations of motion (i.e., all but the inertia coupling) arises from the aerodynamic force and moment.

While these equations for force and moment were derived directly on the basis of simple harmonic oscillations of the airfoil, it is also possible to obtain equivalent results by considering more general unsteady motions of the airfoil. The mathematical analyses are again much too involved to present here, but it is worthwhile to discuss the results briefly. The detailed analyses may be found in References 1, 2, and 8, and those mentioned in subsequent paragraphs of this article.

Consider an airfoil which starts impulsively from rest to a uniform velocity V, and let the motion start at time $t = 0$. The vertical component of velocity of the fluid, the downwash, will be

$$w = V \sin \alpha \simeq V\alpha \tag{5–18}$$

If we define a dimensionless quantity s, measuring the distance traveled by the airfoil in semichords, as

$$s = \frac{Vt}{b} \tag{5–19}$$

and employ the Kutta condition, the lift per unit span, due to the growth of circulation on the airfoil, will be

$$L_1 = 2\pi b\rho V w\varphi(s) \tag{5–20}$$

The function $\varphi(s)$ is known as *Wagner's function* and has the form shown in Fig. 5–3.

Suppose that we consider now that a vertical velocity function is applied to the airfoil at the instant when $s = 0$. The lift resulting from this vertical velocity can be expressed as follows: During a small time increment, the downwash increases by an amount $\frac{dw(s)}{ds}\,ds$; if ds is very small, this may be considered as an impulsive increment of downwash, and the resulting lift per unit span due to circulation may be expressed in terms of a Duhamel integral, as discussed in Chapter 1. Thus,

$$L = 2\pi\rho bV\left[w(0)\varphi(s) + \int_0^s \varphi(s - \bar{s})\,\frac{dw}{d\bar{s}}\,d\bar{s}\right] \tag{5–21}$$

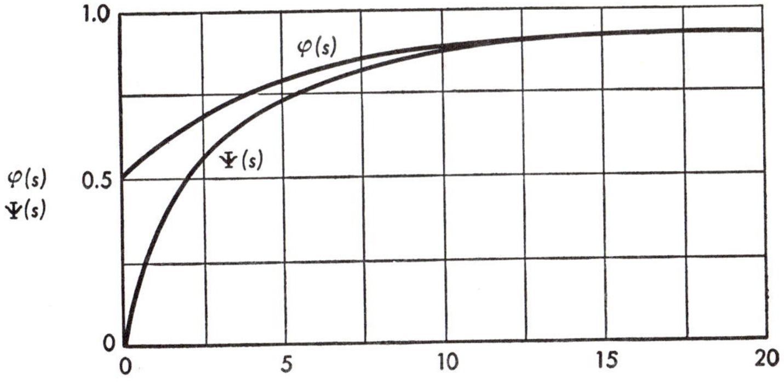

FIG. 5-3. Wagner $\varphi(s)$ function and Küssner $\Psi(s)$ function for unsteady motion.

where $w(0)$ is the value of the vertical velocity at $s = 0$. Now consider the special case where the vertical velocity is oscillatory, so that

$$w(s) = w_0 e^{iks} \tag{5-22}$$

Then, from Eq. 5-21,

$$\frac{L}{2\pi\rho bV} = w(0)\varphi(s) + w_0 ike^{iks} \int_0^s \varphi(\bar{s})e^{-ik\bar{s}}\, d\bar{s} \tag{5-23}$$

This result contains both steady-state and transient components, so that if we consider only the steady-state component, the result becomes

$$\frac{L}{2\pi\rho bV} = w_0 ike^{-iks} \int_0^\infty \varphi(\bar{s})e^{-ik\bar{s}}\, d\bar{s} \tag{5-24}$$

The Theodorsen theory yields a similar expression for the lift due to an oscillating vertical velocity:

$$\frac{L}{2\pi\rho bV} = w_0 e^{iks} C(k) \tag{5-25}$$

Upon comparing Eqs. 5-24 and 5-25,

$$C(k) = ik \int_0^\infty \varphi(\bar{s})e^{ik\bar{s}}\, d\bar{s} \tag{5-26}$$

and since $C(k) = F(k) + iG(k)$, we can also write that

$$\begin{aligned} F(k) &= k \int_0^\infty \varphi(\bar{s}) \sin k\bar{s}\, d\bar{s} \\ G(k) &= k \int_0^\infty [\varphi(\bar{s}) - 1] \cos k\bar{s}\, d\bar{s} \end{aligned} \tag{5-27}$$

Consider now an airfoil entering a sharp-edged vertical gust. If the vertical velocity in the gust is w, then the growth in lift as the airfoil penetrates the gust is given by

$$L = 2\pi\rho bVw\Psi(s) \tag{5–28}$$

where $\Psi(s)$ is known as *Küssner's function.* This function is also shown in Fig. 5–3. The Wagner and Küssner functions are themselves related (Reference 9) according to the expressions

$$\Psi(s) = \frac{1}{\pi}\int_0^s \varphi(s - \bar{s})\sqrt{\frac{\bar{s}}{2 - \bar{s}}}\,d\bar{s} + \frac{1}{\pi}\sqrt{s(2 - s)}, \qquad 0 < s < 2 \tag{5–29a}$$

$$\Psi(s) = \frac{1}{\pi}\int_0^2 \varphi(s - \bar{s})\sqrt{\frac{\bar{s}}{2 - \bar{s}}}\,d\bar{s}, \qquad s > 2 \tag{5–29b}$$

A more general theory of nonuniform airfoil motion has been presented in Reference 10. For rather complete discussions of unsteady-flow theory the student should consult References 1 and 2.

Before closing this discussion of aerodynamic force and moment, it may be worthwhile to outline the derivation for the lift and moment of an airfoil in general motion in terms of the Wagner function.

The downwash at a given point on the airfoil (say, the ¾-chord point*), resulting from the motion of the airfoil, will consist of a uniform component due to the angle of pitch α, a uniform component due to the vertical velocity $\dot{h}$, and a nonuniform component due to the angular velocity $\dot{\alpha}$. Using the previous notation, these three components may be added to give the total downwash as

$$w(s) = V\alpha(s) + \frac{V}{b}h'(s) + \left(\frac{1}{2} - a\right)V\alpha'(s) \tag{5–30}$$

Here, primes denote differentiation with respect to s, which is considered as a nondimensional time, and a locates the c.t. as before. We may again consider impulsive increments of downwash leading to a Duhamel integral as we did in formulating Eq. 5–21. Thus,

$$L_1(s) = 2\pi\rho bV\int_{-\infty}^{s} \varphi(s - \bar{s})\frac{dw}{d\bar{s}}\,d\bar{s} \tag{5–31}$$

Considering the motion to start at $s = 0$, this becomes

$$L_1(s) = 2\pi\rho bV\left[w_0\varphi(s) + \int_0^s \varphi(s - \bar{s})\frac{dw}{d\bar{s}}\,d\bar{s}\right] \tag{5–32}$$

* Küssner showed that the vertical velocity at the ¾-chord point determines the circulation force on the airfoil in oscillating motion. The lift due to circulation acts at the ¼-chord point in steady motion.

and when Eq. 5–30 is introduced, we have

$$L_1(s) = 2\pi\rho b V^2 \int_{-\infty}^{s} \varphi(s - \bar{s}) \left[\alpha'(\bar{s}) + \frac{1}{b} h''(\bar{s}) + \left(\frac{1}{2} - a\right) \alpha''(\bar{s}) \right] d\bar{s} \tag{5-33}$$

Since the airfoil is undergoing a general motion, we must also include the apparent mass forces. There will be a lift force at the mid-chord given by

$$L_2 = \pi\rho V^2(h'' - ab\alpha'') \tag{5-34}$$

and a lift force at the ¾-chord point given by

$$L_3 = \pi\rho b V^2 \alpha' \tag{5-35}$$

The total lift per unit span is then

$$L_T = L_1 + L_2 + L_3 \tag{5-36}$$

and the total moment about the c.t. point will then be

$$M_T = (\tfrac{1}{2} + a)bL_1 + abL_2 - (\tfrac{1}{2} - a)bL_3 + M_a \tag{5-37}$$

where M_a is a nose-down moment equal to the apparent moment of inertia times the angular acceleration and given by

$$M_a = -\frac{\pi\rho b^2 V^2}{8} \alpha'' \tag{5-38}$$

Eqs. 5–36 and 5–37 are completely equivalent to Eqs. 5–14 and 5–15 except that the former apply to general motions while the latter apply to harmonic motions. If one considers the $\varphi(s)$ function for the special case of oscillating motion, then the relation between $\varphi(s)$ and $C(k)$ is given by Eqs. 5–36 and 5–37.

Returning now to Eqs. 5–14 and 5–15, consider simple harmonic motions such that

$$h = h_0 e^{i\omega t}, \quad \alpha = \alpha_0 e^{i(\omega t + \theta)} \tag{5-39}$$

where θ is a phase angle and h_0 and α_0 are the amplitudes of the motion. The aerodynamic force and moment per unit span can then be written in the form

$$P' = \pi\rho b^3 \omega^2 \left\{ L_h \frac{h_0}{b} + \left[L_\alpha - \left(\frac{1}{2} + a\right) L_h \right] \alpha_0 \right\} \tag{5-40}$$

$$M' = \pi\rho b^4 \omega^2 \left\{ \left[M_h - \left(\frac{1}{2} + a\right) L_h \right] \frac{h_0}{b} + \left[M_\alpha - \left(\frac{1}{2} + a\right)(L_\alpha + M_h) + \left(\frac{1}{2} + a\right)^2 L_h \right] \alpha_0 \right\} \tag{5-41}$$

The quantities L_h, L_α, M_h, and M_α depend upon the Theodorsen circulation function $C(k)$ and are often called *flutter coefficients*. This manner of writing the aerodynamic force and moment was first suggested in Reference 11 and has been widely used since. These coefficients have been rather extensively tabulated for the case of flutter involving aileron rotation, in addition to wing torsion and bending; Reference 8 contains such tables. Other schemes have been developed for setting up and tabulating flutter coefficients, but we shall not discuss these here.

5–5. Solution of the Flutter Determinant. The flutter determinant derived in the preceding article is now to be considered. Because the coefficients A, B, C, and D are in general complex quantities, the frequency equation will contain real and imaginary parts, each of which must be zero. Thus, there will be two equations in terms of the two unknowns ω and V. These two equations, although algebraic, are not linear, and therefore they require that rather elaborate methods of solution be employed.

The situation is further complicated because the coefficients of the determinant cannot be set down until the value of the reduced frequency k is known, since the aerodynamic terms are all functions of k. However, k also involves the unknown frequency and velocity; therefore, it appears that the algebraic problem is quite intricate.

If a value of k is chosen, the remaining unknown in the equation will be ω (or V). Since there are two equations, each of which is zero (one from the real part of the frequency equation and one from the imaginary part), one might expect that two different values for ω will be found. The objective, then, is to determine a value for k such that the two values of ω will be equal. Furthermore, the value of ω found by this procedure must be real. The entire process must of necessity be one of trial and error. When compatible values of k and ω have been determined, the flutter velocity V is easily obtained from $k = b\omega/V$.

Theodorsen (Reference 3) has given a simple method for solving the flutter determinant. If the reduced frequency k is assumed, then the determinantal equation may be solved for two unknowns, one of which is the flutter frequency ω. This avoids the necessity for working with two values of ω as described in the preceding paragraph. The other unknown may be taken as some convenient physical parameter of the system (as, for example, the ratio of the uncoupled natural frequencies ω_h and ω_α). Defining,

$$\Omega_\alpha = \left(\frac{\omega_h}{\omega_\alpha}\right)^2 \frac{M_w b^2}{I_w}$$
$$X = \left(\frac{\omega_\alpha}{\omega}\right)^2 \frac{1}{\pi \rho b^2 M_w} \tag{5–42}$$

the frequency equation may be written as a polynomial in the real quantity X, with complex coefficients. This equation may then be separated into real and imaginary parts, each of which must be equal to zero. By solving for roots X_r of the equation from the real part, and roots X_i of the equation from the imaginary part, and plotting the results versus k, the intersection of the two curves thus formed will give the proper values of k and X for the frequency equation to vanish.

In practice, the computations are facilitated if one works with $\sqrt{X}$ and $1/k$. A typical plot of $\sqrt{X}$ versus $1/k$ is shown in Fig. 5–4.

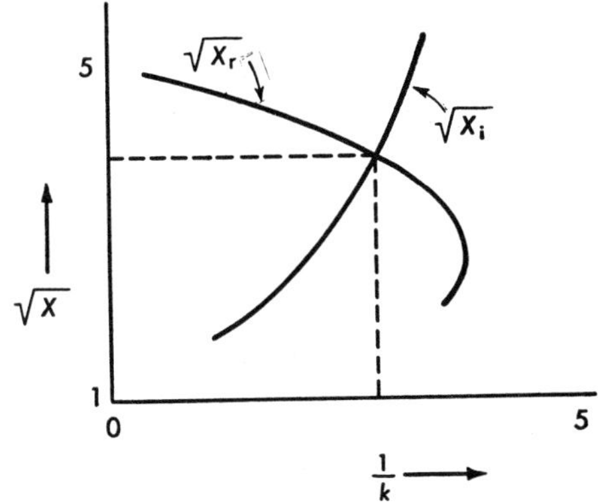

Fig. 5–4. Typical solution of the flutter determinant by the Theodorsen method in terms of the real variable X.

The use of a physical parameter of the system as one of the unknowns is extremely useful. In this manner it is possible to obtain additional information concerning the flutter frequency and velocity of the system. For example, in the case defined by Eq. 5–42, a few different values of Ω_α would show the dependence of flutter frequency (and therefore flutter velocity) on the bending-torsion stiffness (frequency) ratio. Another set of physical quantities which might be employed in this technique would be the ratio of structural damping coefficients, or a function involving wing mass and air density; in the latter case, a variation of flutter speed with altitude may be obtained.

Many different methods have been developed for solving the flutter determinant; some are graphical in nature, some are rather mathematical, and some lend themselves well to tabular or automatic calculation, each method having its own advantages and disadvantages. For a rather simple problem in binary flutter, a graphical method may be satisfactory; for a very complex problem of many degrees of freedom, one of the more elaborate methods of computation may be more desirable. A very detailed account of many of these methods may be found in Reference 8.

5–6. Matrix Formulation of the Flutter Problem. A very interesting and complete formulation of the flutter problem has been given by Loring (References 12 and 13). The entire procedure makes full use of the matrix notation and can therefore be written down with a minimum of effort. Any number of degrees of freedom may be easily accounted for with little additional work, as far as setting up the problem is concerned. Furthermore, matrix notation is extremely convenient for use with automatic computing machines.

According to Loring's analysis a set of "space functions," $\varphi_x, \varphi_y, \varphi_z$, will be used as the generalized coordinates of the problem; these might be, for example, the flexural and torsional mode shapes of a wing. A set of "time functions," q_i, will be used as the generalized displacements; these might be, for example, the flexural and torsional deflections of given points on the wing and are functions only of the time. The generalized displacements are scalars, while the generalized coordinates are vectors.

The displacement of any point P of a structure may then be expressed in terms of the generalized coordinates and generalized displacements, since it will in general be a function of the space variables x, y, z, and of the time t. If u_x, u_y, and u_z are the components of the displacement, we have

$$\begin{aligned} u_x &= \sum_{i=1}^{\infty} q_i\varphi_{xi} = [q]'[\varphi_x] = [\varphi_x]'[q] \\ u_y &= \sum_{i=1}^{\infty} q_i\varphi_{yi} = [q]'[\varphi_y] = [\varphi_y]'[q] \\ u_z &= \sum_{i=1}^{\infty} q_i\varphi_{zi} = [q]'[\varphi_z] = [\varphi_z]'[q] \end{aligned} \tag{5–43}$$

This formulation for displacement is the same as that used before when the displacement of a vibrating beam was given in terms of its normal modes as (see Eq. 2–54)

$$y = \sum_{n=1}^{\infty} (A_n \cos \omega_n t + B_n \sin \omega_n t) \sin \frac{n\pi x}{l}$$

The kinetic energy of a given volume of mass may now be formed:

$$T = \tfrac{1}{2} \iiint (\dot{u}_x^2 + \dot{u}_y^2 + \dot{u}_z^2)\, \bar{\rho}\, dx\, dy\, dz \tag{5–44}$$

where $\bar{\rho}$ is the mass per unit volume of the structure and is a function of x, y, and z. The terms $\dot{u}_x^2$, $\dot{u}_y^2$, $\dot{u}_z^2$ will be of the form

$$\dot{u}_r^2 = [\dot{q}]'[\varphi_r][\varphi_r]'[\dot{q}]$$

since $[q]$ is a column matrix. Then the kinetic energy becomes

$$T = \tfrac{1}{2} [\dot{q}]' \left(\iiint \bar{\rho}([\varphi_x][\varphi_x]' + [\varphi_y][\varphi_y]' + [\varphi_z][\varphi_z]')\, dx\, dy\, dz \right) [\dot{q}] \tag{5–45}$$

The terms within the volume integral all have the dimensions of a moment of inertia and form a square symmetrical matrix with constant elements. Therefore, if the volume integral is called the *inertia matrix* $[a]$, we have

$$T = \tfrac{1}{2}[\dot{q}]'[a][\dot{q}] \tag{5–46}$$

The potential energy may be determined from a volume integral of the strain energy of the structure.* The expressions for the strains may be put in terms of the space derivatives of the displacement components u_x, u_y, and u_z. An expression very similar to Eq. 5–45 will then be obtained, the only difference being that there appears the column matrix of generalized displacements q_i and not its time derivative, and that the terms within the volume integral are space derivatives of the generalized coordinates and elastic moduli such as E and G. Calling the square matrix formed by this volume integral the *stiffness matrix* $[f]$, we have for the potential energy

$$U = \tfrac{1}{2}[q]'[f][q] \tag{5–47}$$

The airforce coefficients of Theodorsen (Reference 3) may be put into matrix form, although the method of doing so is somewhat complicated.† The matrix representing the generalized forces may then be written in terms of an *airforce matrix* $[c]$, whose elements are complex dimensionless functions of the reduced frequency k:

$$Q = -\pi\rho\omega^2 b^4 s[c][q] \tag{5–48}$$

where s is the wing semispan.

These relations (Eqs. 5–46, 5–47, and 5–48) may then be put into Lagrange's equation (see Eq. 2–33)

$$\frac{d}{dt}\left(\frac{\partial T}{\partial \dot{q}_r}\right) - \frac{\partial T}{\partial q_r} + \frac{\partial U}{\partial q_r} = Q_r \tag{5–49}$$

with the result

$$[a][\ddot{q}] + ([f] + \pi\rho\omega^2 b^4 s[c])[q] = [0] \tag{5–50}$$

The generalized airforce matrix is valid only for oscillations of constant amplitude; therefore, solutions of Eq. 5–50 may be taken in the form

$$q = q_0 e^{i\omega t} \tag{5–51}$$

Introducing this expression into the differential equation (Eq. 5–50) will give

$$\left(-\Gamma[a] + [c] + \frac{\Gamma}{\omega^2}[f]\right)[q_0] = [0] \tag{5–52}$$

* See, e.g., S. P. Timoshenko, *Theory of Elasticity* (New York: McGraw-Hill Book Co., Inc., 1934).

† See Ref. 12 for complete details.

where

$$\Gamma = \frac{1}{\pi \rho b^4 s} \tag{5-53}$$

The flutter determinant is then

$$\left| -\Gamma[a] + [c] + \frac{\Gamma}{\omega^2}[f] \right| = 0 \tag{5-54}$$

This brief analysis constitutes a complete *formulation* of the flutter problem. Many details have been omitted, of course, but the analysis is essentially as given. The original papers (References 12 and 13) contain most of the details, as well as extensions and suggestions for tabular methods of computation, and a method for solving the flutter determinant (Eq. 5–54).

Once the flutter determinant has been obtained, its solution for ω and V can be derived by any one of the many methods available. However, it is not very often that the determinantal equation can be *obtained* without considerable effort. The $[a]$ and $[f]$ matrices may, for a complete flutter analysis, often require many weeks of labor for their evaluation.

A complete formulation of the flutter problem by means of matrix analysis has also been presented in Reference 14.

5–7. Effects of Compressibility and Aspect Ratio. As mentioned previously, airplane design for high-speed flight has dictated rather thin wings, leading to fairly flexible structures. This fact, combined with the larger aerodynamic loads associated with high-speed flight has caused the flutter problem to assume ever increasing importance. These higher flight speeds require that the effect of compressibility be taken into account when dealing with the aerodynamic terms in the flutter analysis.

It is well known that the effect of compressibility in the subsonic speed region for steady flow may be accounted for fairly well by the Prandtl-Glauert rule. According to this rule the pressure coefficients, and therefore the forces, for compressible and incompressible flow are related by the expression

$$C_{p_{\text{comp}}} = \frac{C_{p_{\text{inc}}}}{\sqrt{1 - M^2}} \tag{5-55}$$

where M is the free-stream Mach number. Note that this rule predicts an infinite pressure coefficient at $M = 1$, and therefore is not valid for $M \to 1$.

An empirical formula, based on the Prandtl-Glauert rule and introduced by Garrick (Reference 15), relates the compressible and incompressible flutter *velocities* according to

$$V_{\text{comp}} = (1 - M^2)^{1/4} V_{\text{inc}} \tag{5-56}$$

This relation predicts a decrease in flutter speed due to compressibility effects. However, we have seen that the Prandtl-Glauert rule is not valid for $M \rightarrow 1$, and since the relation (Eq. 5–56) is based on that rule, we may say that the empirical formula of Garrick is valid only up to some $M < 1$. Furthermore, since the lift curve slope increases with Mach number, it appears that the lift coefficients would exceed permissible values for some $M < 1$. It is generally agreed that the relation (Eq. 5–56) should not be used for $M > 0.7$. An analysis for the range $0.7 < M < 1$ appears to be a formidable problem because of the presence of shock waves and the nature of the flow, which is partly subsonic and partly supersonic. The study of Reference 15 also indicates that for ordinary wings of normal density and a low bending-torsion frequency ratio, the compressibility correction to the flutter speed is to reduce the flutter speed by only a few per cent.

A more rigorous procedure for study of compressibility effects would be, of course, to derive expressions for an oscillating wing in compressible flow. Discussions of the aerodynamic problem are given in References 1 and 2, while tables of the coefficients are given in References 16–18. Using tables such as those presented in Reference 16, which are in the notation of Smilg and Wasserman (Reference 11), we are able to employ Eqs. 5–40 and 5–41 directly for performing a flutter analysis for a wing in compressible flow.

The analysis for airforces in a purely supersonic flow is somewhat easier. The linearized theory of supersonic flow is not difficult to work with, and results which appear to be quite reasonable have been obtained (Reference 19). The procedure is not unlike that of Theodorsen for the incompressible case; velocity potentials are set up from which the pressures may be obtained; hence, the forces. Extended tables are given in Reference 20.

An attempt to provide airforce coefficients for flow at sonic speed has been given by Nelson and Berman (Reference 21). These authors have taken the linearized theory of supersonic flow and, roughly speaking, extrapolated down to sonic speed while carefully accounting for the unsteady nature of the flow, in which case the linearized theory is applicable. Calculations made by this theory seem to be fairly reasonable when compared with the subsonic and supersonic results as discussed above. For small values of the flutter frequency (i.e., when the case of steady motion is approached) the analysis is inapplicable because of the breakdown of the linearized theory.

A typical variation of bending-torsion flutter velocity V_F with Mach number is shown in Fig. 5–5, with the density factor $m/(\pi\rho b^2)$ as a parameter (Reference 21). Additional parameters are $\omega_h/\omega_\alpha = 0$, $x_\alpha = 0.2$, and $a = 0$, where x_α is the distance from the c.t. to the c.g.

In the flutter analysis of the preceding sections, a representative unit-span portion of the wing was considered. If it is desired to make an analysis where the amplitudes are not to be considered constant along the span, a simple Rayleigh type of analysis may be made. Deformation shapes must be assumed for the wing; these are usually taken to be the

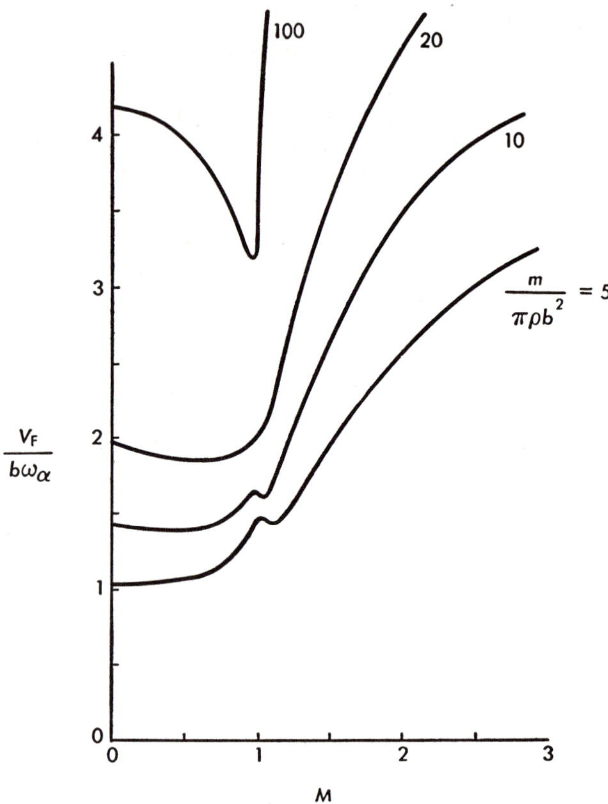

FIG. 5–5. Variations of flutter speed with Mach number, with density ratio as a parameter (Reference 21).

uncoupled torsional and flexural modes of the wing with no airforces acting. Then the equations of motion may be set up with each term in the differential equations expressed as an integral over the span. The principal assumption made here is that the airforces do not change the deflection shapes appreciably. The reasonableness of this assumption will depend, of course, upon the particular structure and problem being analyzed. The aerodynamic forces are treated according to a *strip theory* wherein the local lift coefficient is considered as equal to the product of the local angle of attack and the wing-lift curve slope, the latter being corrected for aspect ratio.

Several methods have been proposed for taking into account the three-dimensional flow over a wing of finite span. Generally, those methods involve the introduction of a correction term or factor to the Theodorsen $C(k)$ function; one such method is given in References 22 and 23. The computations for the effect of finite span are quite complicated and do not always yield reasonable results; therefore, one must use discretion in attempting to apply such theories. Also, one must consider very carefully the question of whether the very great additional effort required for such calculations will be worthwhile for a given problem. Several of the various finite span theories are reviewed and compared in Reference 24.

Some computations incorporating finite span effects seem to show that aspect ratio has a rather large influence on wings of low aspect ratio and low values of k; for high-frequency oscillations, aspect ratio seems to have less effect and is probably negligible for moderate or high aspect-ratio wings.

In general, it appears that the effect of aspect ratio is to raise the flutter speed to a higher value than is given by the two-dimensional analysis (although cases to the contrary have been reported). Some flutter analysts have stated, as a rule of thumb, that for wings of conventional planform the lowering of the flutter speed due to compressibility effects and the raising of the flutter speed due to aspect ratio will very nearly nullify one another. Such "rules" can be applied only after one has had considerable experience in flutter analysis, and even then they must be used with caution.

5–8. Additional Remarks. One of the simpler and often encountered ternary flutter problems is that of bending-torsion of the lifting surface plus the rotation of a control surface about its hinge line (see Fig. 5–6). Actually, the introduction of this additional degree of freedom into the

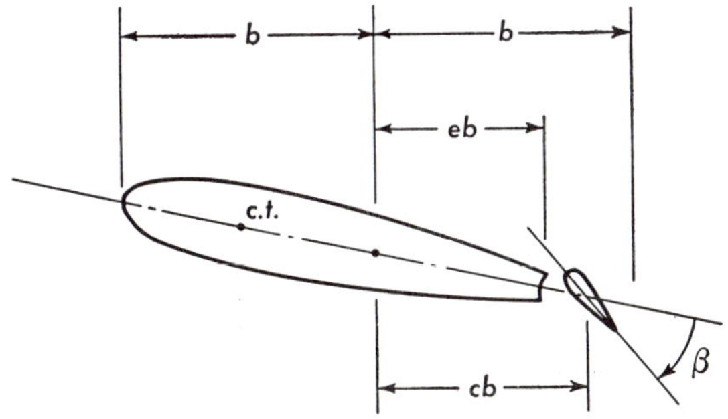

FIG. 5–6. Additional notations for aileron motion about the hinge line.

flutter problem does not result in any undue complication. The airforce coefficients for an airfoil with a hinged control surface are available (References 7, 8, 11, 16–21), and the contributions of the control surface to the kinetic and potential energies are not difficult to evaluate. There will be, of course, three equations of motion to solve, and therefore the flutter determinant will be somewhat more complicated.

It is necessary to distinguish between two different types of control surfaces. Almost all control surfaces are "balanced" in one manner or another in order to reduce the force necessary to provide a given deflection. An *aerodynamically* balanced control surface is so designed that under the action of the airloads, it will be in static equilibrium in every position. An *inertially* balanced control surface is so designed that it will have zero product of inertia about its hinge axis. This type of balancing will prevent any "inertia" coupling between the motion of the lifting surface and the control surface. Hence, by proper balancing, the coupling due to control surface rotation may be eliminated from the flutter problem.

There are three rather obvious approaches to the problem of flutter prevention, any one of which may or may not be sufficient by itself. These are (1) increased structural damping, (2) increased stiffness (especially torsional stiffness), and (3) proper mass-balancing of the wing.

The third method of flutter prevention probably requires some amplification. It has been pointed out previously that there is a certain point on the wing section known as the *center of twist* (see page 87). If the reference point on the airfoil is taken at the center of twist (c.t.), then there will be no elastic coupling between the torsional and flexural motions; there will still be inertia and aerodynamic coupling. The magnitude of the inertia coupling depends upon the product of inertia; therefore, if the wing is mass-balanced about the center of twist, there will be no inertia coupling. The aerodynamic coupling arises from torsional displacements and torsional and flexural accelerations (see Eqs. 5–14 and 5–15). If the center of twist and the aerodynamic center of the wing section were to coincide, then there would be no coupling, due to flexural velocity. Therefore, flutter can be prevented if the center of twist and the aerodynamic center coincide *and* if the wing is mass-balanced about this point.

Early in this chapter it was stated that in order for flutter to exist, it is necessary that coupled bending-torsion vibrations be present. Under certain very specialized conditions (which have, however, been met in practice) it is possible for a surface to flutter in a single degree of freedom —usually torsion. It has been found that in the case of a wing in a supersonic flow, there are certain frequencies and c.t. positions for which the aerodynamic damping is always negative, so that flutter can occur in a pure torsion mode. Some progress has been made on this problem; however, it is as yet inadequately understood. A comparison between theory and experimental results for such a type of flutter is given in Reference 25. (See also Art. 9–1 and Reference 2.)

It seems desirable, before closing this chapter on flutter, to make at least a few remarks concerning the effects of sweepback. A sweptback wing offers a problem in structural analysis of considerable complexity; thus, the determination of the inertia and elastic properties required for the flutter analysis is made much more difficult. Furthermore, the interaction between flexural and torsional motions will be very pronounced, so that not only will the modes of vibration and frequencies be more difficult to calculate than for the straight wing, but also more modes of vibration will probably have to be considered.

Apart from the structural considerations of the sweptback wing, the aerodynamic problem is made very much more difficult by the pronounced three-dimensional character of the flow. As of the present time there appears to be no completely acceptable aerodynamic treatment for the flutter of sweptback wings; Reference 26 gives a summary of some studies concerning flutter of sweptback wings, while Reference 27 presents a method of analysis now in wide use. It appears that in most cases, sweepback will cause some increase in flutter speed,* which may be offset, however, by the effects of compressibility.

Very frequently design requirements dictate that large concentrated weight items be placed along the wing span (as, for example, in the case of engine nacelles and fuel tanks). The presence of such concentrated masses invalidates one of the major assumptions of simple beam theory, and therefore vibration analyses of such wings must include correction terms. The simple beam theory assumes that elements of the beam do not rotate while undergoing transverse vibrations; however, it is clear that large concentrated masses will undergo such rotational motions. These effects are most pronounced in the higher modes and may often lead to serious errors if they are not properly accounted for (see References 8, 28–30).

The formulation of the flutter problem made in Art. 5–3 employed *uncoupled* vibration modes. The contention has often been stated that one would be more realistic if *coupled* vibration modes were used instead. In that representation each vibration mode would be made up of bending and torsion displacement components. The coupled modes will be identical with the uncoupled modes only for the case where the c.g. and

* This is very dependent upon the frequency ratio; if a swept and a straight wing which have the same frequency ratio are compared, the flutter speed of the swept wing will be lower than that of the straight wing.

c.t. exactly coincide at each station along the wing span. When a vibration test of an airplane wing is performed, the mode shapes that are observed are the coupled (free vibration) modes; therefore an excellent method of verifying calculated inertia and elastic properties of the wing by direct experimentation is available. Also, an easier answer may be found to the question, "in what vibration mode does the wing flutter?" Surely, one cannot say that the answer is the second bending mode, or any other such specific motion, simply because the wing does not vibrate in any single independent mode but only in some coupled bending-torsion mode. A rather complete discussion of the use of coupled modes in flutter analysis may be found in Reference 8.

PROBLEMS

5–1. With the aid of Eqs. 5–5, 5–10, 5–14, and 5–15, write out the expressions for A, B, C, and D in the flutter determinant (Eq. 5–12).

5–2. Prove that the expressions for aerodynamic force and moment given by Eqs. 5–36 and 5–37 for an airfoil in a general unsteady motion, when specialized to the case of harmonic motion, are exactly equivalent to the Theodorsen equations (Eqs. 5–14 and 5–15).

5–3. Derive expressions for the flutter coefficients L_h, L_α, M_h, and M_α, as used in Eqs. 5–40 and 5–41.

5–4. An airplane wing has a span of 40 ft. A typical unit span of this wing, having a chord of 5 ft, is estimated to have a weight of 100 lb and a radius of gyration of 2 in. A 20-lb vertical load applied at one end of the wing at the flexural center produces a vertical deflection of 0.10 in.; a pure torque at that point of 20 in.-lb will produce an angular deflection of 0.01 radian. Assume that the flexural and torsional stiffnesses of the wing are constant along the span.

(a) What would you estimate the flutter frequency of this wing to be? Why?

(b) An estimate of the flight flutter velocity, based on your chosen value, turns out to be about 500 mph. However, flight tests later confirm your estimate as to flutter frequency, but the value of the flight flutter velocity was found to be smaller (about 435 mph). How do you explain this lowering of the flutter speed (assuming that the original estimate of 500 mph was based on sound analytical procedures using a bending-torsion analysis and the Theodorsen coefficients)?

5–5. An airplane wing is known to have the following properties:

$$M_w = 0.267 \text{ slug/ft}$$
$$S_w = 0.200 \text{ slug-ft/ft}$$
$$I_w = 2.67 \text{ slug-ft}^2\text{/ft}$$
$$\text{wing semispan} = 15 \text{ ft}$$
$$g_h = g_\alpha = 0.05$$
$$\omega_\alpha = 1100 \text{ cycles/min}$$
$$\omega_h = 800 \text{ cycles/min.}$$

At a certain representative station on the span we have the following values: $b = 2.5$ ft; $ab = -1.0$ ft. Using these given values, expand the flutter determinant

(Eq. 5–12). Using the method outlined on pages 98–99, determine the flutter speed and flutter frequency.

5–6. Using the notation of Figs. 5–1 and 5–6, derive equations corresponding to Eq. 5–10 for the combination of wing bending, wing torsion, and aileron rotation about the hinge line for a representative wing section of unit span.

REFERENCES

1. Fung, Y. C. *The Theory of Aeroelasticity.* New York: John Wiley & Sons, Inc., 1955.
2. Bisplinghoff, R. L., Ashley, H., and Halfman, R. L. *Aeroelasticity.* Reading, Mass.: Addison-Wesley Publishing Co., Inc., 1955.
3. Theodorsen, T. "General Theory of Aerodynamic Instability and the Mechanism of Flutter," *NACA Rept.* 496 (1935).
4. Theodorsen, T., and Garrick, I. E. "Mechanism of Flutter—A Theoretical and Experimental Investigation of the Flutter Problem," *NACA Rept.* 685 (1940).
5. Coleman, R. P. "Damping Formulas and Experimental Values of Damping in Flutter Models," *NACA Tech. Note* 751 (1940).
6. Soroka, W. W. "Note on the Relations Between Viscous and Structural Damping Coefficients," *Jour. Aero. Sciences,* **16** (July, 1949): 409–410.
7. Luke, Y., and Dengler, M. "Tables of the Theodorsen Circulation Function for Generalized Motion," *Jour. Aero. Sciences,* **18** (July, 1951): 478–483.
8. Scanlan, R. H., and Rosenbaum, R. *Introduction to the Study of Aircraft Vibration and Flutter.* New York: The Macmillan Co., 1951. *(Dover reprint)*
9. Garrick, I. E. "On Some Reciprocal Relations in the Theory of Nonstationary Flows," *NACA Rept.* 629 (1938).
10. Kármán, Th. von, and Sears, W. R. "Airfoil Theory for Non-Uniform Motion," *Jour. Aero. Sciences*; **5** (October, 1938): 379–390.
11. Smilg, B., and Wasserman, L. S. "Application of Three-Dimensional Flutter Theory to Aircraft Structures," *AAF Tech. Rept.* 4798 (July, 1942).
12. Loring, S. J. "General Approach to the Flutter Problem," *S.A.E. Journal,* **49** (August, 1941): 345–355.
13. Loring, S. J. "Use of Generalized Coordinates in Flutter Analysis," *S.A.E. Journal,* **52** (April, 1944): 113–132.
14. Barndollar, E. J. "Application of Matrix Operations to Flutter Analysis," *Douglas Aircraft Company Rept. SM*–13693 (1950).
15. Garrick, I. E. "Bending-Torsion Flutter Calculations Modified by Subsonic Compressibility Corrections," *NACA Rept.* 836 (1946).
16. Luke, Y. L. "Tables of Coefficients for Compressible Flutter Calculations," *Air Force Tech. Rept.* 6200 (1950).
17. Timmen, R., Van de Vooren, A. I., and Greidanus, J. H. "Aerodynamic Coefficients of an Oscillating Airfoil in Two-Dimensional Subsonic Flow," *Jour. Aero. Sciences,* **18** (December, 1951): 797–802. (See discussions and corrections in the same journal, 1952–1954.)
18. Blanch, G. "Tables of Lift and Moment Coefficients for Oscillating Airfoils in Subsonic Compressible Flow;" *National Bureau of Standards Rept.* 2260 (1953).
19. Garrick, I. E., and Rubinow, S. I. "Flutter and Oscillating Air-Force Calculations for an Airfoil in a Two-Dimensional Supersonic Flow," *NACA Rept.* 846 (1946).
20. Huckel, V., and Durling, B. "Tables of Wing-Aileron Coefficients of Oscillating Air Forces for Two-Dimensional Supersonic Flow," *NACA Tech. Note* 2055 (1950).
21. Nelson, H. C., and Berman, J. H. "Calculations on the Forces and Moments for an Oscillating Wing-Aileron Combination in Two-Dimensional Potential Flow at Sonic Speeds," *NACA Rept.* 1128 (1953).
22. Reissner, E. "Effect of Finite Span on the Airload Distributions for Oscillating Wings I," *NACA Tech. Note* 1194 (1947).
23. Reissner, E., and Stevens, J. E. "Effect of Finite Span on the Airload Distributions for Oscillating Wings II," *NACA Tech. Note* 1195 (1947).

24. WASSERMAN, L. S. "Aspect Ratio Corrections in Flutter Calculations," *USAF, AMC Memo. Rept. MCREXA* 5–4995–8–5 (August, 1948).
25. RUNYAN, H. L. "Single-Degree-of-Freedom-Flutter Calculations for a Wing in Subsonic Potential Flow and Comparison with an Experiment," *NACA Rept.* 1087 (1952).
26. BARMBY, J. G., CUNNINGHAM, H. J., and GARRICK, I. E. "Study of Effects of Sweep on the Flutter of Cantilever Wings," *NACA Rept.* 1014 (1951).
27. SPIELBERG, I., FETTIS, H. E., and TONEY, H. S. "Method for Calculating the Flutter and Vibration Characteristics of Swept Wings," *USAF, AMC Memo. Rept. MCREXA* 5–4595–8–4 (August, 1948).
28. SCANLAN, R. H. "A Note on Transverse Bending of Beams Having Both Translating and Rotating Mass Elements," *Jour. Aero. Sciences*, **15** (July, 1948): 425–426.
29. FETTIS, H. E. "Effect of Rotary Inertia on Higher Modes of Vibration," *Jour. Aero Sciences*, **16** (July, 1949): 445.
30. ANDERSON, R. A., and SEWALL, J. L. "Experimental Investigation of the Effects of Concentrated Weights on Flutter Characteristics of a Straight Cantilever Wing," *NACA Tech. Note* 1594 (1948).

CHAPTER 6

AEROELASTICITY

6–1. Introduction. *Aeroelasticity* may be defined as that branch of the aeronautical science which considers the interaction between aerodynamic, elastic, and inertia forces on aircraft components during flight. A striking example of structural deformation during flight is shown in Fig. 6–1. A characteristic common to all aeroelastic problems is that the

Boeing Airplane Co.

Fig. 6–1. Deformation of XB-52 wing during normal flight.

aerodynamic loads give rise to structural deformation. These deformations, in turn, cause changes in the aerodynamic loads, which again change the deformation, and this process repeats until a state of equilibrium is reached. The aerodynamic forces, however, do not respond instantaneously to structural deformation but instead "lag behind." This lag, called *aerodynamic lag,* is accounted for in the Theodorsen analysis of flutter by the circulation terms. Thus, we see that aeroelasticity represents a combined problem in structural and aerodynamic analyses.

Of all the aeroelastic problems, probably the three most important ones are flutter, wing torsional divergence, and reduction in control surface effectiveness. The first of these, which is a dynamics problem, was discussed in an elementary manner in Chapter 5; the remaining two, which are essentially static problems, will be discussed in the subsequent articles of this chapter. Then there will be a brief discussion of the effects of structural deformation on the general aerodynamic characteristics, and finally an elementary derivation of the equations of aeroelasticity will be presented.

There are yet other problems which might well be classified as aeroelastic problems—among them are the problems of buffeting, gust response, landing impact, and aeroelastic effects on flight stability. The first three of these have the common feature that they are all concerned with the dynamic response of structural components to a given external excitation. This excitation may be purely aerodynamic in origin, as in the case of buffeting and gust response, or it may originate as a sudden blow or shock to the structure, as in the case of landing impact. These "response" problems will be discussed in Chapter 7. A few remarks concerning aeroelastic effects on flight stability will be found in Chapter 8 which deals with the general problem of flight stability.

The above account of the general picture of airplane dynamics is, of course, far from complete. For example, not mentioned are the problems of propeller and rotor vibration, the dynamics of shock-absorbing systems of all types, engine vibrations, etc. A few general remarks concerning such topics will be found in Chapter 9.

6–2. Static Wing-Torsional Divergence. Wing divergence, although definitely an aeroelastic problem, is not a problem of vibration. Instead, it is purely a static problem in which the stability of the structure under the applied loads is of prime concern. More precisely, wing divergence refers to the condition at which the destabilizing aerodynamic moment exceeds the elastic restoring moment in torsion, thereby causing large torsional displacements of the wing.

Let us consider for a moment the total torsional stiffness of a wing. The elasticity of the structure provides part of this stiffness, and since it always tends to restore the wing structure to its equilibrium configuration, we may speak of the torsional rigidity of the structure as providing a positive torsional stiffness. The remaining contribution to torsional stiffness is aerodynamic in origin, and while the elastic contribution is a constant for the wing under all conditions, the aerodynamic contribution may be positive or negative.

We could say, then, that divergence of a wing occurs when its total torsional stiffness becomes negative or, in the limiting case, passes from positive to negative.

To understand better the action of the aerodynamic contribution to torsional stiffness, consider a wing which has been slightly twisted. The twist causes an additional lift by virtue of the change in angle of attack. This additional lift acts through the aerodynamic center (a.c.),* so that if the a.c. is *aft* of the center of twist, the additional lift will tend to decrease the twist; it therefore contributes a positive torsional stiffness. If, however, the a.c. lies *forward* of the center of twist, as is more usually the case with actual wings, then the additional lift contributes a negative torsional stiffness.

It is well known that aerodynamic forces increase with the square of the flight velocity; therefore, it is clear that there is some flight velocity at which the negative torsional stiffness would be exactly equal to the positive torsional stiffness (contributed by the torsional rigidity of the structure)—this is the *speed of torsional divergence.*

Inasmuch as this is a static problem, the formulation is quite easy and may be made directly from the equations of Chapter 5. From Eqs. 5–8 and 5–15 we may write, by simply omitting all dynamic terms,

$$C_\alpha \alpha = \pi\rho V_d^2 b^2(1 + 2a)\alpha \tag{6–1}$$

Now, by introducing the uncoupled natural frequency in torsion, we may obtain a very simple expression for the divergence speed:

$$V_d^2 = \frac{\omega_\alpha^2 I_\alpha}{\pi\rho b^2(1 + 2a)} \tag{6–2}$$

This result applies to a representative unit-span section of a wing and may be considered as approximating the actual wing-divergence speed. However, a much better value may be obtained, and the wing divergence mode as well, in a very simple manner by use of matrix methods.

Consider the wing span as divided into strips of width Δy, so that for the ith strip Eq. 6–1 becomes, with $\lambda = \pi\rho V^2$,

$$\Delta C_{\alpha_i}\alpha_i = \lambda b_i^2(1 + 2a_i)\Delta y_i \alpha_i \tag{6–3}$$

If we define a diagonal matrix $[Y]$ whose elements are $(1 + 2a_i)b_i^2\,\Delta y_i$, we have

$$\{\Delta C_\alpha \alpha\} = \lambda[Y]\{\alpha\} \tag{6–4}$$

Now, we can define a matrix $[C]$ according to

$$[C]\{\Delta C_\alpha \alpha\} = \{\alpha\} \tag{6–5}$$

and combine Eqs. 6–4 and 6–5 so that

$$\{\alpha\} = \lambda[K]\{\alpha\} \tag{6–6}$$

* J. H. Dwinnell, *Principles of Aerodynamics* (New York: McGraw-Hill Book Co., Inc., 1949), p. 114. The lift due to camber (as produced by the deflection of an aileron, for example) acts much farther aft on the airfoil.

where

$$[K] = [C][Y] \tag{6-7}$$

Eq. 6–6 may now be solved by matrix iteration; the lowest value of λ yields the wing-divergence speed, and the corresponding modal column of $\{\alpha\}$ describes the divergence mode.

An even more simple analysis can be given if the wing is idealized to an unswept cantilever, and we assume that a straight spanwise line, normal to the fuselage, can be passed through all the centers of twist (such a line is called the *elastic axis*). Denoting the distance from the line of aerodynamic centers to the elastic axis as $2b\sigma$ (positive aft), the aerodynamic moment at a given station y_i about the elastic axis due to a change in angle of attack (θ_i) is

$$M_i \, \Delta y_i = 4\bar{a}\rho V^2 \sigma_i b_i^2 \theta_i \, \Delta y_i \tag{6-8}$$

where $\bar{a}$ is the lift curve slope corrected for aspect ratio.

At the critical divergence speed, this aerodynamic moment must be exactly balanced by the elastic restoring force provided by the structure. Equating these two moments in differential equation form, we have

$$\frac{d}{dy}\left[GJ\frac{d\theta}{dy}\right] = -4\bar{a}\rho V_d^2 \sigma(y) b^2(y) \theta(y) \tag{6-9}$$

where GJ represents the torsional stiffness of the wing as a function of the span. This differential equation may be solved if the wing geometry and structure are known, although the solution may be very difficult in many cases. The associated boundary conditions are

$$\begin{aligned} \theta &= 0, \quad (y = 0) \\ \frac{d\theta}{dy} &= 0, \quad (y = s) \end{aligned} \tag{6-10}$$

where s is the wing semispan. Consider the special case of a uniform rectangular wing so that Eq. 6–9 becomes

$$\frac{d^2\theta}{dy^2} = -\beta^2\theta \tag{6-11}$$

where

$$\beta^2 = \frac{4\bar{a}V_d^2\sigma b^2}{GJ} \tag{6-12}$$

The general solution of this differential equation is

$$\theta = A \sin \beta y + B \cos \beta y \tag{6-13}$$

The boundary conditions (Eq. 6–10) require that

$$B = 0, \quad \cos \beta s = 0 \tag{6–14}$$

so that for the lowest value of β,

$$\beta = \frac{\pi}{2s} \tag{6–15}$$

Thus,

$$V_d^2 = \frac{GJ\pi^2}{4\bar{a}\rho\sigma b^2 s} \tag{6–16}$$

The corresponding divergence mode of the wing is

$$\theta(y) = \theta_0 \sin \frac{\pi y}{2s} \tag{6–17}$$

These analyses, although quite simple, should still serve to indicate essential points of the problem. It may be specifically mentioned, however, that for low aspect ratio wings, wings with appreciable sweep, or wing structures not easily defined in terms of simple conceptions such as the elastic axis, much more refined analyses are required in both the aerodynamic and structural aspects of the analysis. For example, the structural deformation may have to be described in terms of influence coefficients due to unit loads and the aerodynamic forces by a complex lifting-surface theory.

The type of compressibility correction applied to the flutter speed in Chapter 5 (Eq. 5–56) may be also applied here to the divergence speed. Therefore, the effect of compressibility is to decrease the divergence speed.

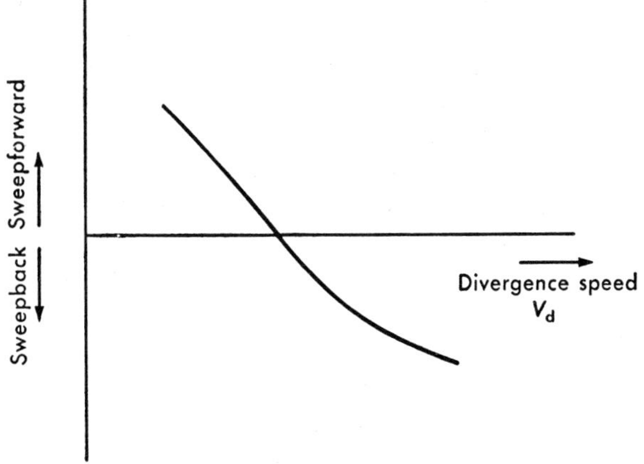

FIG. 6–2. Effect of sweep on divergence speed.

The very pronounced coupling between flexural and torsional displacements which exists in swept wings would seem to indicate that the divergence speed would depend upon the amount of sweep. In fact, it is shown in Reference 3 that sweepforward causes a large reduction in the divergence speed, while sweepback causes a large increase (Fig. 6–2).

Further aspects of the divergence problem and discussions of more accurate methods are given in References 1–10.*

6–3. Control Surface Reversal. The phenomenon of control-surface reversal may be described as a reversal of the net normal force on the lifting surface, induced by the deflected control surface, due to the aerodynamic moment twisting the elastic structure.

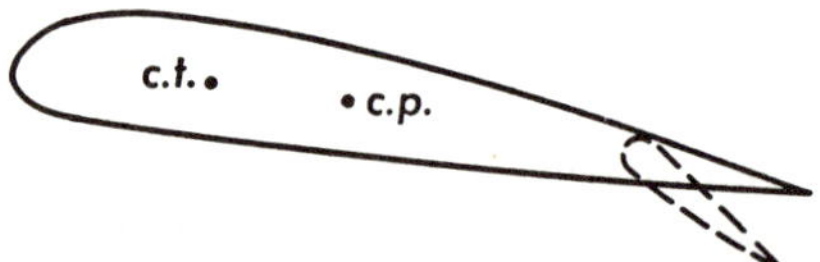

FIG. 6–3. Typical airfoil section for discussion of divergence and reversal.

Assume that we have a wing with an aileron deflected down (Fig. 6–3) and with the center of pressure aft of the center of twist. The increment of lift due to the aileron deflection will be such as to twist the wing so that the angle of attack is decreased; thus, the total lift on the wing is somewhat reduced. It is clear, then, that the elasticity of the structure causes a reduction in the effectiveness of the control surface. The wing stiffness in torsion is always a constant, but the aerodynamic force varies with the square of the velocity; therefore, it appears that there is some flight speed at which the effect of the aileron deflection will be completely nullified by the distortion of the wing. At some higher flight velocity, the distortion effect will be so great that the aileron deflection will produce an effect opposite to that desired. The flight speed at which the effectiveness of the control surface is zero is known as the *control reversal speed.*

Inasmuch as the ailerons are used primarily as a control for the rolling moment of a wing, it is clear that the rolling effectiveness of a given wing will be markedly affected by the presence of any significant amount of the torsional flexibility.

A very simple formula for the calculation of the aileron reversal speed has been advanced by Pugsley (Reference 11):

$$V_r = K \sqrt{\frac{C_\alpha}{2\rho b^2 s}} \tag{6–18}$$

* See Art. 9-1 for a few brief remarks concerning the divergence of supersonic wings.

where K is a factor dependent upon the aerodynamics and geometry of the wing-aileron combination, and s is the wing semispan.

A somewhat better analysis, based on strip theory, may be derived similarly to the divergence analysis of the preceding article. From the Theodorsen analysis of the harmonically oscillating wing-aileron combination (Reference 12), the nonoscillating lift per unit span can be expressed as

$$P = -2\pi\rho b V^2 \left[\alpha + \left(1 - \frac{\varphi}{\pi} + \frac{1}{\pi}\sin\varphi\right)\beta\right] \tag{6–19a}$$

$$= -2\pi\rho b V^2 \left(\alpha + \frac{T_{10}}{\pi}\beta\right) \tag{6–19b}$$

where

$$x = -\cos\varphi \tag{6–20}$$

defines any given point on the airfoil measured from the mid-chord, and β is the angular deflection of the control surface about its hinge line relative to the wing chord (stalling angle positive).

If we consider that the aileron is so constructed that it twists with the wing, i.e., β is a constant along the wing span, we may write the wing rolling moment as

$$M_{\text{roll}} = -4\pi\rho V^2 \left[\sum_{\text{wing}} (b_i y_i \alpha_i \,\Delta y_i) + \sum_{\text{ail}} \left(\beta b_i y_i \frac{T_{10}}{\pi}\Delta y_i\right)\right] \tag{6–21}$$

At the reversal speed, the wing rolling moment is zero, and therefore we may solve for β. Now, the nonoscillating moment about the elastic axis for a strip Δy_i can be written as

$$\Delta M_i = \pi\rho V^2 \left\{ (1 + 2a)_i b_i^2 \alpha_i + \left(2a\frac{T_{10}}{\pi} - \frac{T_4}{\pi}\right)_i b_i^2 \beta \right\} \Delta y_i \tag{6–22}$$

where

$$T_4 = -\pi + \varphi - \sin\varphi\cos\varphi \tag{6–23}$$

The angle β may be eliminated between Eqs. 6–21 and 6–22 for the case of zero wing rolling moment, and the moment can then be written in the form

$$\Delta M_i = \pi\rho V^2 \sum_{k=1}^{n} F_{ik}\alpha_k \tag{6–24}$$

or in matrix notation

$$\{\Delta M\} = \lambda[F]\{\alpha\} \tag{6–25}$$

Note that the elements of the matrix $[F]$ are functions of T_4 and T_{10}, which in turn depend on the aileron parameter e (see Fig. 5–6). Now, if we introduce the matrix of torsional influence coefficients $[C]$, as defined in Eq. 6–5, we have

$$\{\alpha\} = \lambda[G]\{\alpha\} \tag{6–26}$$

where $[G] = [C][F]$. This matrix equation may be solved by the iteration method, as before, to find the reversal speed from the lowest value of λ.

A more definite expression may be obtained in the following manner: We note from Eq. 6–19b that the condition for reversal may be expressed in the form (omitting second-order terms)

$$\frac{dP}{d\beta} = -2\pi\rho b V^2 \left(\frac{\partial\alpha}{\partial\beta} + \frac{T_{10}}{\pi}\right) = 0 \tag{6–27}$$

or

$$\frac{\partial\alpha}{\partial\beta} = -\frac{T_{10}}{\pi} \tag{6–28}$$

The aerodynamic pitching moment per unit span can be written, from Eq. 6–22, as

$$M = \pi\rho b^2 V^2 \left\{ (1 + 2a)\alpha + \left(2a\frac{T_{10}}{\pi} - \frac{T_4}{\pi}\right)\beta \right\} \tag{6–29}$$

and this is to be balanced by the elastic restoring moment of the wing structure, $C_\alpha\alpha$. Again differentiating with respect to β, we find

$$C_\alpha \frac{\partial\alpha}{\partial\beta} = \pi\rho b^2 V^2 \left\{ (1 + 2a)\frac{\partial\alpha}{\partial\beta} + \left(2a\frac{T_{10}}{\pi} - \frac{T_4}{\pi}\right) \right\} \tag{6–30}$$

and eliminating $\partial\alpha/\partial\beta$ by use of Eq. 6–28 then yields

$$V_r^2 = \frac{C_\alpha}{\pi\rho b^2}\left(\frac{T_{10}}{T_{10} + T_4}\right) \tag{6–31}$$

It is to be noted that the reversal speed is directly proportional to the wing-torsional stiffness and does not depend upon the location of the center of twist, in general agreement with the semiempirical formula (Eq. 6–18).

Eq. 6–31 is sometimes expressed in more conventional aerodynamic terms as

$$V_r^2 = \frac{C_\alpha}{2\bar{a}\rho b^2}\left(\frac{\partial C_l/\partial\beta}{-\partial C_m/\partial\beta}\right) \tag{6–32}$$

where the differential lift and moment coefficients are relatively well-known quantities in aerodynamic theory (Reference 13).

The effect of compressibility on the reversal speed may again be accounted for by the Garrick formula, Eq. 5–56. In general, the effect of sweep is to lower the aileron reversal speed, as shown in Fig. 6–4. A very complete discussion of control reversal effects on swept wings is given in Reference 14.

For most aircraft, the center of pressure of the additional lift due to control surface deflection is aft of the elastic axis; in such a case, the reversal speed is usually less than the divergence speed, so that sufficient torsional rigidity to prevent reversal will also prevent divergence. Other studies have shown that an elastic axis position forward of the mid-chord point will generally make the reversal speed lower than the divergence speed. In the case of swept wings, however, the divergence speed is the more critical.

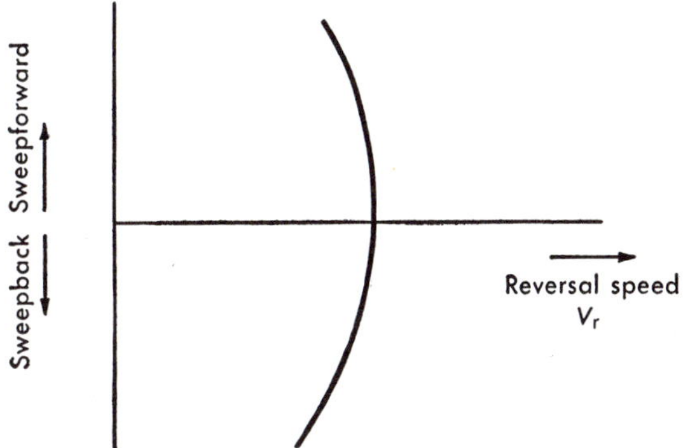

FIG. 6–4. Effect of sweep on reversal speed.

More refined analyses of the control reversal problem may be found particularly in References 1, 2, 8–10, and 15. A set of charts useful in design are given in References 16 and 17.

6–4. Effects of Structural Deformation on Aerodynamic Characteristics. As mentioned previously, the effects of structural deformation on the aerodynamic characteristics are very pronounced, and as may be inferred from the preceding article, the major effect is to reduce the effectiveness of control surfaces. The detrimental effect of wing flexibility on lateral control is shown qualitatively in Fig. 6–5, where the rolling effectiveness of a wing is plotted against flight velocity. The effectiveness of yawing and pitching controls (rudder and elevators) is similarly reduced by aeroelastic effects, since the moments contributed by these control surfaces may cause deformations of the entire aft end of the fuselage.

For a more quantitative statement of the loss of control effectiveness, let us define an *aileron efficiency factor* as the ratio of the lift produced by a unit aileron deflection on an elastic wing to that produced by the same aileron deflection on a similar rigid wing. From Eqs. 6–27 and 6–30,

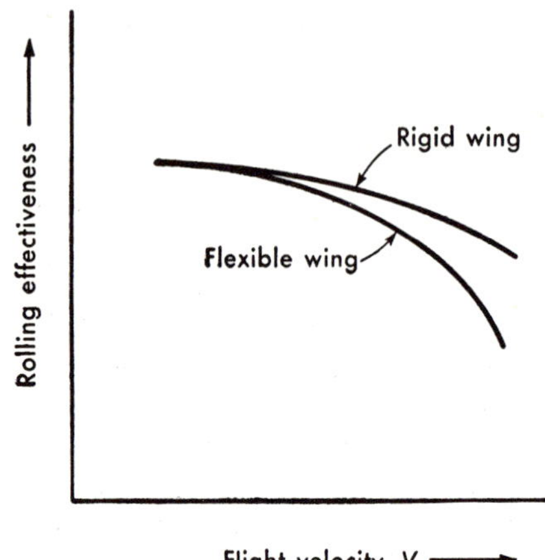

FIG. 6–5. Effect of wing flexibility on rolling (aileron) effectiveness.

the change in lift of an elastic wing with aileron deflection may be written as

$$\frac{dP}{d\beta} = -2\pi\rho b V^2 \left[\frac{T_{10}}{\pi} + \frac{\pi\rho b^2 V^2 \left(2a \frac{T_{10}}{\pi} - \frac{T_4}{\pi} \right)}{C_\alpha - \pi\rho b^2 V^2 (1 + 2a)} \right] \tag{6–33}$$

For a perfectly rigid wing $C_\alpha \to \infty$, so that we have

$$\left(\frac{dP}{d\beta} \right)_{\text{rigid}} = -2\rho b V^2 T_{10} \tag{6–34}$$

The aileron efficiency factor then becomes

$$\frac{dP/d\beta}{(dP/d\beta)_{\text{rigid}}} = 1 + \frac{\pi\rho b^2 V^2 \left(2a - \frac{T_4}{T_{10}} \right)}{C_\alpha - \pi\rho b^2 V^2 (1 + 2a)} \tag{6–35}$$

This result may be put into a very simple form if we introduce the divergence and reversal speeds from Eqs. 6–1 and 6–31. Thus,

$$\frac{dP/d\beta}{(dP/d\beta)_{\text{rigid}}} = \frac{1 - (V/V_r)^2}{1 - (V/V_d)^2} \tag{6–36}$$

We may consider here only the case of $V_r < V_d$, otherwise the aileron efficiency is negative, indicating that the reversal condition exists when $V \geq V_d$. We note also that the case $V_d = V_r$ represents the optimum condition, since the aileron efficiency remains unchanged up to the critical speed. For $V_r < V_d$, the aileron efficiency drops with increasing speed (Fig. 6–5). The problem of rolling (aileron) effectiveness is discussed in detail in References 15–20.

Probably next in importance to the reduction in control effectiveness is the change in span load distribution. Generally, the effect is to reduce the load at the tips and increase the load near the root. This will, of course, rather markedly affect such aerodynamic quantities as wing-lift curve slope, aerodynamic center location, and drag characteristics. One might say, in general, that the effect of the structural deformation is to cause a change in the angle of attack at all stations along the span of the wing; this new distribution of angle of attack along the span will cause an entirely new set of aerodynamic characteristics to be present.

The problem of determining the lift distribution on an elastic wing is one of considerable complexity; for swept wings and for all wings of low aspect ratio the problem is very difficult. The structural deformation is often treated by the influence coefficient method because the elastic axis concept is not applicable for most wings of practical significance in this type of problem. The aerodynamic portion of the problem may be treated in several ways, one of the most simple of which is to define the angle of attack at a station y as composed of the sum of a contribution from the rigid wing section and a contribution from the elastic deformation; thus,

$$\alpha(y) = \alpha_r + \alpha_e \tag{6–37}$$

The lift distribution on the flexible wing can be obtained by a process of successive approximation wherein the rigid wing-lift distribution is first calculated, and then the resulting elastic deformation is determined and used to calculate a new lift distribution. The elastic contribution to the new lift distribution is then used to calculate the new structural deformation, and so on. The rigid wing-lift distribution may be determined by application of a strip theory, by lifting-line theory, or by any of the other well-known aerodynamic methods. Detailed discussions and methods are given in References 6–10, 21–23.

Structural deformations may have great effect in many other cases. For example, a control surface itself may deform torsionally, thereby complicating the reversal problem discussed in the preceding article. Furthermore, the hinge-moment coefficients will no longer be linear functions, a condition which then complicates the entire problem of providing adequate control. Also, major structural deformations, such as vertical bending of the fuselage, play an important role in the flight stability problem. Some general comments on this last problem will be given in Chapter 8.

6–5. The Elementary Equations of Aeroelasticity. It is possible to write equations which govern the aeroelastic problem, after making certain simplifying idealizations. Because the aeroelastic problem concerns the entire wing, some account of finite span must be made; this is usually accomplished by dividing the wing into a certain number of sections or strips and assuming that within each strip the aerodynamic quantities are constant, as was done previously. The aerodynamic effects of the free tips may be accounted for by applying an over-all correction factor (see Art. 5–7). We shall give a very simple analysis for the basic aeroelastic equations in the present article (Reference 7).

The discussion of aeroelasticity thus far has demonstrated the importance of the center of twist of the wing section. If we now speak of a wing of finite span, it will be necessary to refer to some sort of axis in the wing about which the twisting action takes place. Such an axis is called an *elastic axis* (Art. 6–2) and is defined as the locus of the centers of twist* along the span. Thus, the wing is idealized to a single line having at each section the bending and torsional rigidity of the corresponding section of the actual wing.

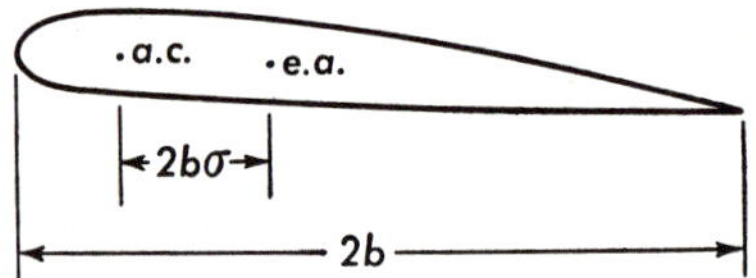

FIG. 6–6. Section notation for aeroelasticity analysis.

It is usually assumed that the elastic axis is a straight line which intersects the root section at right angles. This assumption will enable us to consider all deformations about the elastic axis to be given by the elementary theories of bending and torsion, as was done in the divergence calculation of Art. 6–2. As usual, small deflections will be assumed.

The lift force (per unit width) on a wing section perpendicular to the elastic axis may be written as

$$P = q(2b)(C_{L_\alpha}\alpha_r + C_{L_{\alpha e}}\alpha_e) \tag{6–38}$$

where

α_e = angle of attack due to structural deformation
α_r = rigid wing angle of attack
C_{L_α} = wing-lift curve slope
$C_{L_{\alpha e}}$ = effective wing-lift curve slope†
q = dynamic pressure

and the wing is assumed to have no wash-out or wash-in.

* Again we do not distinguish between the center of twist, shear center, and flexural center in this simplified discussion.

† The effective lift curve slope is used for lift distributions due to aeroelastic twist, roll, etc. An approximate value is given in Reference 7 as simply a correction factor (smaller than unity) applied to C_{L_α}. Also see Art. 6–4.

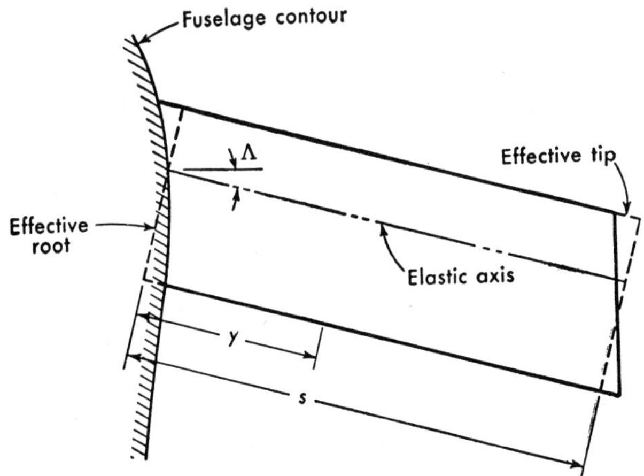

FIG. 6-7. Wing notation for aeroelasticity analysis.

The torque produced by this lift force about the elastic axis is $P(2b\sigma)$, where σ is the dimensionless distance from the aerodynamic center to the elastic axis (see Figs. 6-6 and 6-7 for the wing geometry). The total torque and moment at a station y along the span are then

$$T = \int_y^s P(2b\sigma)\, dy \tag{6-39}$$

$$M = \int_y^s \int_y^s P\, dy\, dy \tag{6-40}$$

where y is the spanwise coordinate measured along the elastic axis and s is the wing semispan.

The torsional and flexural displacements are measured as follows: Torsional deflection α is measured in a plane perpendicular to the elastic axis. Flexural deflections φ are measured (see Fig. 6-8) as a dihedral angle, i.e., as the spanwise slope of the normal displacement of the elastic

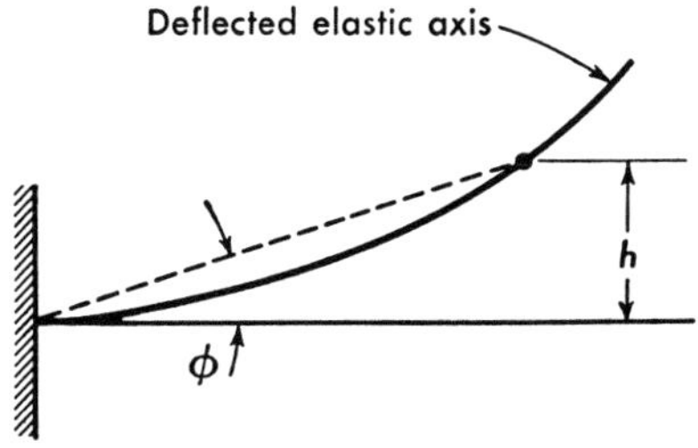

FIG. 6-8. Notation for aeroelasticity analysis.

axis. The deflections at a point y on the span produced by the torque and moment are then

$$\alpha = \int_0^y \frac{T}{GJ}\,dy \tag{6–41}$$

$$\varphi = \int_0^y \frac{M}{EI}\,dy \tag{6–42}$$

The angle of attack due to structural deformation α_e is related to the deflections α and φ by means of the sweepback angle Λ:

$$\alpha_e = \alpha \cos \Lambda - \varphi \sin \Lambda \tag{6–43}$$

Differentiating Eqs. 6–41 and 6–42, introducing Eqs. 6–39 and 6–40, differentiating once again, and then introducing Eq. 6–43 gives

$$\frac{d}{dy}\left(GJ\,\frac{d\alpha}{dy}\right) = -q\sigma(2b)^2[C_{L_\alpha}\alpha_r + C_{L_{\alpha e}}(\alpha \cos \Lambda - \varphi \sin \Lambda)] \tag{6–44}$$

$$\frac{d^2}{dy^2}\left(EI\,\frac{d^2\varphi}{dy^2}\right) = q(2b)[C_{L_\alpha}\alpha_r + C_{L_{\alpha e}}(\alpha \cos \Lambda - \varphi \sin \Lambda)] \tag{6–45}$$

Eqs. 6–44 and 6–45 represent the fundamental differential equations of aeroelasticity. Note that these equations are very similar to the usual moment equation for beams, in which the right-hand sides represent the coupled moment loading (compare with Eq. 2–55a). The boundary conditions which may be used to solve these equations are as follows:

At the Root ($y = 0$)

$$\alpha = 0, \quad \varphi = 0 \tag{6–46}$$

because the wing is idealized to the extent of being rigidly clamped at the root.

At the Tip ($y = s$)

$$\begin{aligned} GJ\,\frac{d\alpha}{dy} &= 0 \quad \text{(Torque)} \\ EI\,\frac{d\varphi}{dy} &= 0 \quad \text{(Moment)} \\ EI\,\frac{d^2\varphi}{dy^2} &= 0 \quad \text{(Shear)} \end{aligned} \tag{6–47}$$

Solutions of these fundamental equations for several cases are given in Reference 7. The results of such computations have been presented in a series of charts and formulas in References 7 and 17 which are extremely useful for design purposes. This analysis is quite simple and convenient, but it must be recalled that the elastic axis concept, and hence this analysis, is not valid for many modern wing structures.

PROBLEMS

6–1. A certain airplane is observed to exhibit a slight tendency toward torsional divergence at a speed of 485 mph. It is desired to increase this divergence speed by 10 per cent. How would you accomplish this? (Give a numerical answer.)

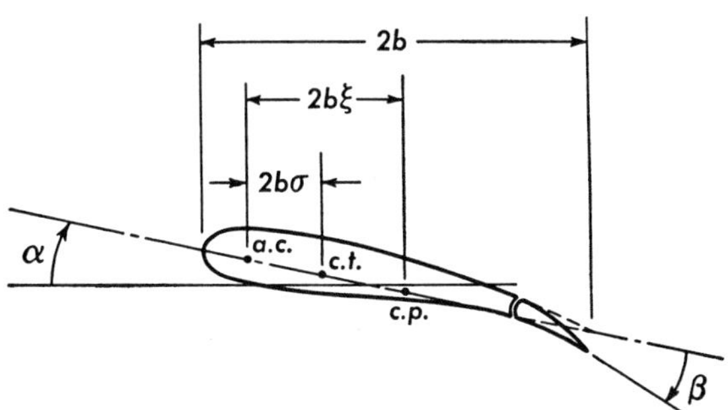

FIG. 6–9. Problem 2.

6–2. Consider a simple rectangular airfoil with an aileron as shown in Fig. 6–9. Let the additional lift due to the twist of the airfoil (α) and the deflection of the aileron (β) be given by

$$L_a = c_1\alpha + c_2\beta$$

where c_1 and c_2 are constants. The effectiveness of the aileron may be defined as the ratio of the additional lift (as given above) to the additional lift produced by the aileron alone. Using this information, derive the expression given by Pugsley (Eq. 6–18) for the aileron reversal speed. Discuss the expression which you obtain for K.

6–3. Explain why the control reversal speed is independent of the location of the center of twist.

6–4. For the wing described in Problem 5 of Chapter 5, calculate the divergence and control reversal speeds (use $k = 0.2$).

REFERENCES

1. FLAX, A. H. "The Influence of Structural Deformation on Airplane Characteristics," *Jour. Aero. Sciences*, **12** (January, 1945): 94–102.
2. FLAX, A. H. "Aeroelastic Problems at Supersonic Speeds," *Proc. Second Int. Aero. Conf.*, New York (May, 1949), 322–360.
3. DIEDERICH, F. W., and BUDIANSKY, B. "Divergence of Swept Wings," *NACA Tech. Note* 1680 (August, 1948).
4. COLLAR, A. R. "Aeroelastic Problems at High Speed," *Jour. Roy. Aero. Soc.*, **51** (January, 1947): 1–23.
5. PAI, S. I., and SEARS, W. R. "Some Aeroelastic Properties of Swept Wings," *Jour. Aero. Sciences*, **16** (February, 1949): 105–115.
6. BROWN, R. B., HOLTBY, K. F., and MARTIN, H. C. "A Superposition Method for Calculating the Aeroelastic Behavior of Swept Wings," *Jour. Aero. Sciences*, **18** (August, 1951): 531–542.
7. DIEDERICH, F. W., and FOSS, K. A. "Charts and Approximate Formulas for the Estimation of Aeroelastic Effects on the Loading of Swept and Unswept Wings," *NACA Rept.* 1140 (1953).

8. SCANLAN, R. H., and ROSENBAUM, R. *Introduction to the Study of Aircraft Vibration and Flutter*. New York: The Macmillan Co., 1951. *(Dover reprint)*
9. FUNG, Y. C. *The Theory of Aeroelasticity*. New York: John Wiley & Sons, Inc., 1955.
10. BISPLINGHOFF, R. L., ASHLEY, H., and HALFMAN, R. L. *Aeroelasticity*. Reading, Mass.: Addison-Wesley Publishing Co., Inc., 1955.
11. PUGSLEY, A. G. "Control Surface and Wing Stability Problems," *Jour. Royal Aero. Society*, **41** (November, 1937): 975–996.
12. THEODORSEN, T. "General Theory of Aerodynamic Instability and the Mechanism of Flutter," *NACA Rept.* 496 (1935).
13. PERKINS, C. D., and HAGE, R. E. *Airplane Performance, Stability and Control*. New York: John Wiley & Sons, Inc., 1949.
14. TEMPLETON, H. "Control Reversal Effects on Swept-Back Wings," *The Aero. Quart.*, **1** (May, 1949): 3–34.
15. HARMON, S. M. "Determination of the Effect of Wing Flexibility on Lateral Maneuverability and a Comparison of Calculated Rolling Effectiveness with Flight Results," *NACA Wartime Rept. L*–525.
16. PEARSON, H. A., and AIKEN, W. J. "Charts for the Determination of Wing Torsional Stiffness Required for Specified Rolling Characteristics or Aileron Reversal Speed," *NACA Rept.* 799 (1944).
17. FOSS, K. A., and DIEDERICH, F. W. "Charts and Approximate Formulas for the Estimation of Aeroelastic Effects on the Lateral Control of Swept Wings," *NACA Rept.* 1139 (1953).
18. HEDGEPETH, J. M., and KELL, R. J. "Rolling Effectiveness and Aileron Effectiveness of Rectangular Wings at Supersonic Speeds;" *NACA Tech. Note* 3067 (1954).
19. DIEDERICH, F. W. "Calculation of the Lateral Control of Swept and Unswept Flexible Wings of Arbitrary Stiffness," *NACA Rept.* 1024 (1951).
20. HEDGEPETH, J. M., WANER, P. G., and KELL, R. J. "A Simplified Method for Calculating Aeroelastic Effects on the Roll of Aircraft," *NACA Tech. Note* 3370 (1955).
21. DIEDERICH, F. W. "Calculation of the Aerodynamic Loading of Swept and Unswept Flexible Wings of Arbitrary Stiffness," *NACA Rept.* 1000 (1950).
22. DIEDERICH, F. W. "A Simple Approximate Method for Calculating Spanwise Lift Distributions and Aerodynamic Influence Coefficients at Subsonic Speeds," *NACA Tech. Note* 2751 (1952).
23. GRAY, W. L., and SCHENK, K. M. "A Method for Calculating the Subsonic Steady-State Loading on an Airplane with a Wing of Arbitrary Planform and Stiffness," *NACA Tech. Note* 3030 (1953).
24. BISPLINGHOFF, R. L. "Some Structural and Aeroelastic Considerations of High-Speed Flight," *Jour. Aero. Sciences*, **23** (April, 1956): 289–329.

CHAPTER 7

PROBLEMS OF IMPULSIVE LOADING—LANDING IMPACT, GUST RESPONSE, AND BUFFETING

7-1. Introduction. Aircraft structures are very often subjected to high-intensity loads which act over short periods of time. If the time duration of the loading is sufficiently small, the structure is said to have been subjected to an *impact* or an *impulse* type of load. During the impulse, the various particles in the system receive large accelerations, and hence definite changes in velocity are produced, after which the system vibrates about its equilibrium position. An analysis of this type of problem was made in Chapter 1, where a single-degree-of-freedom system was subjected to an external exciting force acting as an arbitrary function of time. The exciting force was considered to be composed of a large number of impulses, and the final response was obtained by summing the responses to these successive impulses. Thus, it was found that the response (particular solution) of the undamped single-degree-of-freedom system governed by the equation of motion

$$\ddot{x} + \omega^2 x = q(t) \tag{7-1}$$

was given by the Duhamel integral

$$x_p = \frac{1}{\omega}\int_0^t q(t')\sin\omega\,(t - t')\,dt' \tag{7-2}$$

It seems reasonable to expect that a similar analysis for a system of particles would again lead to Duhamel's integral; we shall see that an elementary theory of transients in elastic systems does, in fact, lead to this integral.

Strictly speaking, the real transient motion of an elastic system, as discussed in previous chapters, is the motion given by the solution of the homogeneous differential equation; i.e., the free vibrations (which soon disappear due to the damping that always exists in any real system). However, when dealing with impulsive motions, where the maximum response to the external force occurs very early in the motion, it has become customary to call this motion (particular solution of the non-homogeneous equation) the *transient* response—the expression "dynamic" response would seem to provide a much better description of the motion we are discussing.

The vibrations resulting from impulsive loading may at times cause stresses that exceed the design strength of the structure, as evidenced by the spectacular failure, due to an excessive landing impact, shown in the Frontispiece of this book. The possibility of such failures has long been recognized and has become a problem of major importance with the advent of large high-speed aircraft and missiles embodying relatively flexible structural components. Consequently, much effort has been directed toward developing methods for analyzing adequately the dynamic response of complex structures subjected to impulsive loads.

Impulsive or impact loading of aircraft structures may occur in many different ways; three of the more common are landing impact, gust loading, and buffeting. Each of these will be discussed briefly in the subsequent articles of this chapter after an elementary treatment of the theory of dynamic response in elastic systems.

7–2. Elementary Theory of Dynamic Response in Elastic Systems. We are concerned now with the response of an elastic structure to impulsive loading, in which it is usual for the response to reach a maximum value very soon after the motion starts; hence, omission of the effects of damping is justifiable. Inasmuch as we deal with an elastic system, which has an infinite number of degrees of freedom, we may consider the general motion to be made up of a superposition of natural vibration modes, as was found to be the case when studying the vibrations of beams (Chapter 2). An exact treatment would require that an infinite number of vibration modes be considered; however, for most problems, only a few of the lowest modes will be required for a solution of acceptable accuracy. In the analysis of beam vibrations it was found that the formulation of the problem was very clear and useful when expressed in terms of the *normal* modes of vibration; the present analysis will be set forth in terms of such modes.

The normal mode shapes will be represented by a function of space variables φ_n, and the generalized coordinates, which are functions of time, will be denoted by q_n. The displacement of any point in the elastic structure will be given by*

$$u = \sum_n \varphi_n q_n \tag{7–3}$$

The kinetic energy of the structure is easily formulated as

$$T = \frac{1}{2} \int \left(\sum_n \varphi_n \dot{q}_n \right)^2 dm \tag{7–4}$$

* See Art. 5–6.

The potential energy may be written down in simple form by expressing the stiffness in terms of the mass and frequency as

$$U = \tfrac{1}{2} \sum_n M_n \omega_n^2 q_n^2 \tag{7-5}$$

These expressions for the kinetic and potential energies may now be substituted into Lagrange's equation:

$$\frac{d}{dt}\left(\frac{\partial T}{\partial \dot{q}_n}\right) - \frac{\partial T}{\partial q_n} + \frac{\partial U}{\partial q_n} = Q_n(t) \tag{7-6}$$

which holds for each of the generalized coordinates q_n. In this form of Lagrange's equation, $Q_n(t)$ represents the generalized force corresponding to the nth generalized coordinate. We consider that the external force is a function of time only.

Introducing first the kinetic energy gives

$$\frac{d}{dt}\left(\frac{\partial T}{\partial \dot{q}_n}\right) = \ddot{q}_n \int (\varphi_n)^2 \, dm$$

It is a property of the natural vibration modes of elastic systems that

$$\sum_n \varphi_r \varphi_s = \delta_{rs} \tag{7-7}$$

where

$$\begin{aligned} \delta_{rs} &= 1 \quad (r = s) \\ &= 0 \quad (r \neq s) \end{aligned} \tag{7-8}$$

This statement is known as the *orthogonality relation* for the principal modes of vibration—it has already been discussed briefly in Chapters 2 and 3, and additional information is presented in Appendix B. By virtue of the orthogonality property, the inertia term then becomes

$$\frac{d}{dt}\left(\frac{\partial T}{\partial \dot{q}_n}\right) = \ddot{q}_n \int (\varphi_n)^2 \, dm = \ddot{q}_n M_n \tag{7-9}$$

Upon introducing the potential energy, the complete equation of motion for each of the natural modes becomes

$$\ddot{q}_n + \omega_n^2 q_n = \frac{Q_n(t)}{M_n} \tag{7-10}$$

We now see that, as is usual for normal modes, each of the modes behaves as though it were an independent single-degree-of-freedom system. Now, if δW_n is the work produced when the external force is allowed to move through an infinitesimal displacement δq_n, then

$$\delta W_n = \delta q_n Q_n(t) \tag{7-11}$$

from which the generalized force $Q_n(t)$ may be obtained.

The solution of Eq. 7–10 may be written in terms of Duhamel's integral as

$$q_n = \frac{1}{M_n\omega_n}\int_0^t Q_n(t') \sin \omega_n(t - t')\, dt' \tag{7–12}$$

Now, we define a *dynamic load factor* D_n as

$$D_n = \omega_n \int_0^t Q_n(t') \sin \omega_n(t - t')\, dt' \tag{7–13}$$

so that the generalized coordinate in each normal mode can be written

$$q_n = \frac{D_n}{M_n\omega_n^2} \tag{7–14}$$

The displacement, in accordance with Eq. 7–3, can finally be written as

$$u = \sum_n \frac{D_n}{M_n\omega_n^2}\varphi_n \tag{7–15}$$

From this formulation we see rather easily that the time variation of the dynamic response is given entirely by the dynamic load factor. Because of the manner in which the potential energy was formulated, it is also immediately obvious that for statically applied loads the value of D_n is unity. On the other hand, the value of D_n in the case of a load applied suddenly and then held constant with time (see Fig. 7–1) is 2.0. Loading conditions between these two extremes will result in a value of dynamic load factor between 1.0 and 2.0.

The stress in the structure at a point k, due to the deformation in the nth mode, is given by

$$(\sigma_n)_k = (A_n)_k q_n \tag{7–16}$$

where $(A_n)_k$ is the amplitude constant of the nth mode at the point k. It follows that the total stress at the point k is obtained as the super-

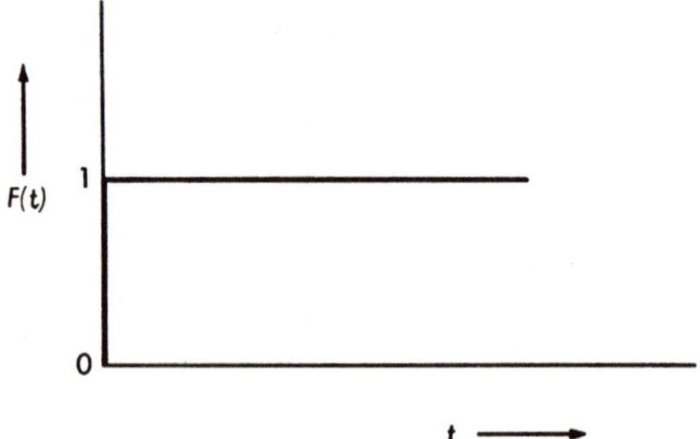

FIG. 7–1. Unit-step function.

position of the stresses produced by the deformations of all the modes, so that

$$\sigma_k = \sum_n (A_n)_k q_n = \sum_n (A_n)_k \frac{D_n}{M_n \omega_n^2} \tag{7-17}$$

As an example of this theory, let us now consider the dynamic response of a simply supported, uniform beam (Fig. 7–2) subjected to unit-step function $F(t)$, as shown in Fig. 7–1, applied at the mid-span. The free

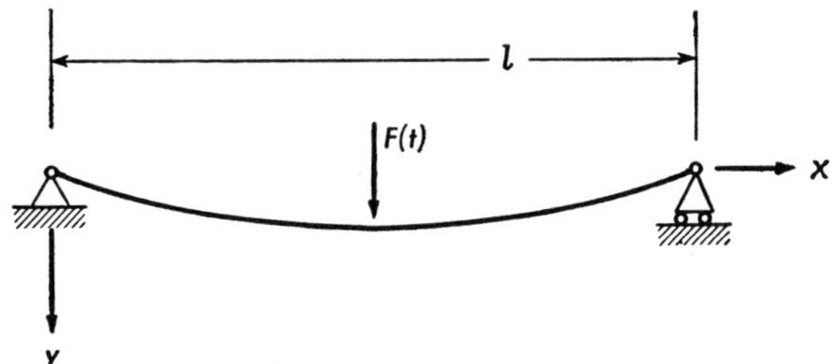

FIG. 7–2. Simply supported beam subjected to an impulsive load at the mid-span.

vibrations of such a beam were discussed in Chapter 2, and it was found that the deflection at a given point on the beam can be written as

$$y = \sum_n A_n \sin \frac{n\pi x}{l} \tag{7-18}$$

or, in the notation of this chapter,

$$y = \sum_n q_n \varphi_n \tag{7-19}$$

It follows that q_n and A_n are corresponding quantities, and thus the generalized coordinates A_n may be expressed in the form of Eq. 7–12. Now

$$M_n = \int (\varphi_n)^2 \, dm = \frac{w}{g} \int_0^l \sin^2 \frac{n\pi x}{l} \, dx = \frac{w}{g} \frac{l}{2} \tag{7-20}$$

and

$$Q_n(t) = (\varphi_n)_F F(t) = \left(\sin \frac{n\pi}{2} \right) F(t) \tag{7-21}$$

The quantity $(\varphi_n)_F$ may be regarded as the normal mode function evaluated at the point of application of the force F. Now, introducing these expressions into Eq. 7–18 gives

$$y = \frac{2g}{wl} \sum_n \frac{1}{\omega_n} \sin \frac{n\pi}{2} \sin \frac{n\pi x}{l} \int_0^t F(t') \sin \omega_n (t - t') \, dt' \tag{7-22}$$

For $F(t)$ as shown in Fig. 7–1, i.e., a unit-step function, the integration may be performed and the dynamic response written as

$$y = \frac{2g}{wl} \sum_n \frac{1}{\omega_n^2} \sin \frac{n\pi}{2} \sin \frac{n\pi x}{l} (1 - \cos \omega_n t) \tag{7–23}$$

Numerical computations have been carried out (Reference 1), based on the foregoing type of analysis, which show that very good results are obtained for the bending moment by considering only the first two modes, and that reasonably good values are obtained for the shear by considering the first four modes. For more severe types of impulsive loading, accurate results can be obtained only if a very large number of modes are considered; if the impulse is extremely sharp, certain of the assumptions of simple beam theory, which we have employed here, may be violated so that a more exact theory may be required (Reference 2).

Under many circumstances the externally applied load acts not only as a function of time but also as a function of the space variables. Therefore, let us extend the preceding analysis to the case where the loading is distributed along the beam. The governing differential equation is

$$EI \frac{\partial^4 y}{\partial x^4} = -\frac{w}{g} \frac{\partial^2 y}{\partial t^2} - F(x, t) \tag{7–24}$$

and the known normal mode equations for the displacement (from Chapter 2) are

$$\frac{d^4 \varphi_n}{dx^4} - \beta_n^4 \varphi_n = 0 \tag{7–25a}$$

$$\varphi_n = A \sin \beta_n x + B \cos \beta_n x + C \sinh \beta_n x + D \cosh \beta_n x \tag{7–25b}$$

$$\beta_n^4 = \frac{\omega_n^2}{EI} \frac{w}{g} \tag{7–25c}$$

Again we suppose that the displacement can be given in terms of a mode shape φ_n and a generalized coordinate q_n, as in Eq. 7–3, and we assume that the applied force can be represented in a similar manner, so that

$$y(x, t) = \sum_n q_n(t)\varphi_n(x) \tag{7–26}$$

$$F(x, t) = \sum_n p_n(t)\varphi_n(x) \tag{7–27}$$

Multiplying both sides of these expressions by $\varphi_n(x)\,dx$ and integrating, and employing the orthogonality condition, we find

$$q_n(t) = \int_0^l y(x, t)\varphi_n(x)\,dx \tag{7–28}$$

$$p_n(t) = \int_0^l F(x, t)\varphi_n(x)\,dx \tag{7–29}$$

Substituting Eqs. 7–26 and 7–27 into Eq. 7–24 gives

$$\sum_n \left[q_n(t) \frac{d^4\varphi_n}{dx^4} + \frac{w}{g} \frac{\varphi_n}{EI} \frac{d^2q_n}{dt^2} + \frac{\varphi_n}{EI} p_n(t) \right] = 0 \tag{7–30}$$

Now, with Eq. 7–25a we may eliminate the term involving $d^4\varphi_n/dx^4$ and find that each term involves $\varphi_n(x)$ as a factor. Multiplying again by $\varphi_n(x)\,dx$ and integrating, we have

$$\frac{d^2q_n}{dt^2} - \omega_n^2 q_n = \frac{g}{wl} p_n \tag{7–31}$$

or, introducing Eq. 7–29,

$$\frac{d^2q_n}{dt^2} - \omega_n^2 q_n = \frac{g}{wl} \int_0^l F(x, t)\varphi_n(x)\,dx \tag{7–32}$$

We shall now assume that the applied force may also be expressed as the product of a space function $f_s(x)$ and a time function $f_t(t)$, according to

$$F(x, t) = F_0 f_s(x) f_t(t) \tag{7–33}$$

and where F_0 is the magnitude of the applied force. The right-hand side of Eq. 7–32 can now be written as

$$\begin{aligned} \frac{g}{wl} \int_0^l F(x, t)\varphi_n(x)\,dx &= \frac{gF_0}{wl} f_t(t) \int_0^l f_s(x)\varphi_n(x)\,dx \\ &= \frac{gF_0}{wl} f_t(t) K_n \end{aligned} \tag{7–34}$$

The quantity

$$K_n = \int_0^l f_s(x)\varphi_n(x)\,dx \tag{7–35}$$

determines the contribution of the nth mode to the motion, and hence it is called the *participation factor.*

The differential equation (Eq. 7–32) can now be solved in terms of Duhamel's integral to give

$$q_n = \frac{gF_0}{wl} \frac{K_n}{\omega_n} \int_0^t f_t(t') \sin \omega_n(t - t')\,dt' \tag{7–36}$$

or, in terms of the dynamic load factor,

$$q_n = \frac{gF_0}{wl} \frac{K_n}{\omega_n^2} D_n \tag{7–37}$$

The beam dynamic response can now be written in terms of the normal mode shapes $\varphi_n(x)$, which will depend in turn upon the boundary conditions (end conditions), as

$$y(x, t) = \frac{gF_0}{wl} \sum_n \frac{K_n}{\omega_n^2} D_n \varphi_n(x) \tag{7-38}$$

The participation factor K_n indicates the extent to which the various modes contribute to the response and depends upon the normal mode shapes and the space distribution of the applied force, while the time variation of the response is given entirely by the dynamic load factor D_n.

A comprehensive study of impact effects on simple structures has been given in Reference 3, while a more recent and general theory of the dynamic response of elastic systems by the normal mode procedure is given in Reference 4.

It has been found possible to analyze the response of an elastic structure to dynamic loads without resorting to the use of modal shapes. The modal type of analysis requires that the generalized orthogonal vibration modes and frequencies of the structure be computed in advance, only the lowest few of these being required because otherwise the actual computation becomes too complex. In many problems, particularly those of impulsive nature, many modes are required in order to describe adequately the response of the system, and hence other methods have been developed to overcome this difficulty.

In Reference 5 there is given a method which does not involve the modal functions. Instead, a set of difference-differential equations are formulated which will enable one to compute the structural response to applied loads. Thus, the procedure is one of step-by-step computation with the requirement that conditions of equilibrium are satisfied at given points in the structure at all times. As an aid to computation and formulation, the entire analysis of Reference 5 is given in matrix form. A recurrence formula is given which is used to evaluate the special form of Duhamel's integral which appears in the analysis.

7–3. Landing Impact. The study of the stresses in an aircraft structure produced by a landing impact may be broken down into two rather distinct problems. The first is to determine the nature of the load that is transmitted to the structure, and the second is to determine the dynamic response of the structure to that load. The physical process involves, first of all, a transformation of the kinetic energy of the airplane into internal energy in the shock-absorption system of the landing gear. This internal energy is then partly dissipated and partly transferred back to the airplane in terms of strain energy by means of the landing-gear reaction or by the acceleration of points of attachment of the landing gear.

In many cases, however, interaction between the motions of the landing gear and the deformations of the airplane structure causes a modification of the behavior of the landing gear. The trend toward larger and more flexible airplanes, often with large masses concentrated at outboard locations on the wings, have resulted in reductions in the natural vibration frequencies, with a consequent increase in the amplitudes of motion of gear-attachment points relative to the center of gravity of the airplane. Thus, the use of load time histories, without consideration for the effects of airplane flexibility on the behavior of the landing, may often be far from adequate in predicting the inertia loads and stresses in the airplane structure.

The conventional oleo-pneumatic landing gear of today presents a difficult problem in dynamics, not only because of its inherently complex motion and function, but also because of the numerous external factors which affect its behavior. Some of the more important of these external factors are, for example, the nonlinear force-deflection characteristics of the tire, the drag load resulting from wheel spin-up, the lift and velocity of the airplane at touchdown, the inclination and orientation of the landing gear, the condition and type of runway, and the flexibility of the structure to which the gear is attached.

Generally speaking, the analysis of a landing gear proceeds as follows: The actual mechanical system must be idealized to a spring-mass-damper system of one or more degrees of freedom.* Next, the expressions which account for the internal forces of the system, such as forces due to air compression and internal friction in the shock strut, and the orifice discharge coefficient, as well as the external factors noted above, must be formulated. The differential equations of the system may then be derived. For any but the most simplified analysis, these equations will not be linear, and therefore they cannot be solved by the methods used in earlier chapters; instead, one must resort to numerical methods of integration (References 7 and 8).

As a very simple example of a landing-gear system, let us consider a rigid airplane mass m_1 connected to a landing gear m_2 by a linear spring k_1, and a linear dash pot c; the tire is assumed to be a linear spring k_2. The system is shown schematically in Fig. 7–3 at the time $t = 0$, which represents the instant at which the tire makes contact with the ground.† The displacements of the masses m_1 and m_2, measured from their positions at $t = 0$, are called x_1 and x_2, respectively. The airplane mass is subjected to two forces: one is the weight of the airplane, slightly reduced to account for a portion of the wing lift, and denoted by W, while the other is the

* A very simple idealization of an airplane landing-gear combination to a two-degree-of-freedom system was shown in Fig. 2–11.

† We consider here only a vertical descent with velocity V.

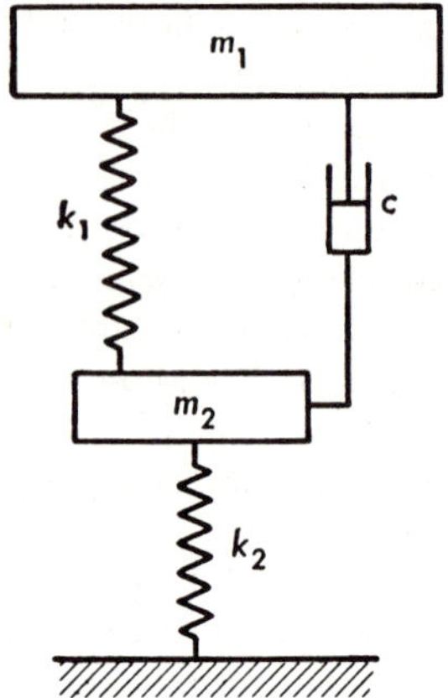

FIG. 7–3. Idealized airplane landing-gear system.

force in the shock strut which depends upon the spring k_1 and the damper c. The differential equation of motion for the airplane mass then becomes

$$m_1\ddot{x}_1 = W - [k_1(x_1 - x_2) + c(\dot{x}_1 - \dot{x}_2)] \tag{7–39a}$$

The landing-gear mass is also subjected to two forces: one is again the force in the shock strut, and the other is the force transmitted from the ground by the tire, so that the differential equation of motion becomes

$$m_2\ddot{x}_2 = [k_1(x_1 - x_2) + c(\dot{x}_1 - \dot{x}_2)] - k_2x_2 \tag{7–39b}$$

These two differential equations describe the landing-gear impact under the highly idealized conditions described. The solutions of these equations may be written as

$$x_1 = W\frac{k_1 + k_2}{k_1k_2} + e^{-\alpha_1 t}(A_1 \cos \omega_1 t + A_2 \sin \omega_1 t) + e^{-\alpha_2 t}(A_3 \cos \omega_2 t + A_4 \sin \omega_2 t) \tag{7–40a}$$

$$x_2 = \frac{W}{k_2} + e^{-\alpha_1 t}(B_1 \cos \omega_1 t + B_2 \sin \omega_1 t) + e^{-\alpha_2 t}(B_3 \cos \omega_2 t + B_4 \sin \omega_2 t) \tag{7–40b}$$

In these solutions α_1, α_2, ω_1, and ω_2 all depend upon physical constants of the system and are obtained from the frequency equation. The A's and B's are directly related to each other by physical constants of the system, so that there remains either the set of A's or the set of B's as arbitrary constants to be determined by the initial conditions. For the problem described, these initial conditions would be

$$x_1 = 0, \quad \dot{x}_1 = V$$
$$x_2 = 0, \quad \dot{x}_2 = V \tag{7-41}$$

This simple model may now be expanded by the addition of various terms to account more accurately for the operation of an actual landing-gear system; such analyses are given in Reference 8. Other simplified models, considering varying degrees of structural flexibility, are presented in Reference 9, with special reference to the stresses developed in the wing structure.

We wish now to consider the second part of the over-all problem which refers to the dynamic response of the airplane structure to the applied loads. Much of the work in this area has been based on the concept of the natural vibration modes of an elastic structure, particularly beams, as discussed in the preceding article (see References 1, 10–13). Detailed computation methods for such analyses are given in References 10 and 13.

A systematic procedure for estimating on a statistical basis the dynamic stresses in airplane structures during landing was developed some time ago (Reference 1). The stresses corresponding to the maximum displacements in a few of the lowest modes are first added, without regard to phase differences. The maximum amplitude of stress in each mode is then estimated from an envelope of "dynamic stress factors," determined from a statistical survey, which indicates the response to any force-time relationship that may be expected in the landing. This simple type of analysis is not applicable to most modern aircraft configurations and has largely been replaced by more rigorous techniques (Reference 14).

Methods that do not depend on modal descriptions have also been developed (References 5 and 15). A comparison of such methods with the modal methods is given in Reference 16, and some general comments will also be given at the end of this chapter.

We come now to a brief consideration of some of the effects of the interaction between the flexibility of the airplane structure and the landing gear (References 17 and 18). The effects of interaction may be discussed first from the point of view of modifying the dynamic loads in the airplane structure.

Because of the flexibility of the structure, the landing-gear attachment points may move with appreciable amplitudes relative to the airplane center of gravity. Thus, the downward velocity of the shock-strut outer cylinder is more rapidly dissipated, and its displacement is therefore reduced throughout most of the impact. The tire deflection is also smaller, but because of its relatively great stiffness, this decrease is smaller than that of the shock-strut outer cylinder. The net result is a reduction in the stroke during the part of the impact when the applied force is greatest, and there is therefore an accompanying reduction in the velocity of compression of the shock strut. Since the maximum landing-gear force is primarily due to the hydraulic portion of the system, the reduction in stroke and velocity of compression leads to a decrease in the shock-strut force. The amount of reduction of the shock-strut force depends also upon the ratio of the elastically supported mass to the mass acting directly on the landing gear, as well as upon the design characteristics of the particular landing gear.

The modification of the dynamic loads in the airplane structure may consist of either dynamic amplification or attenuation compared with

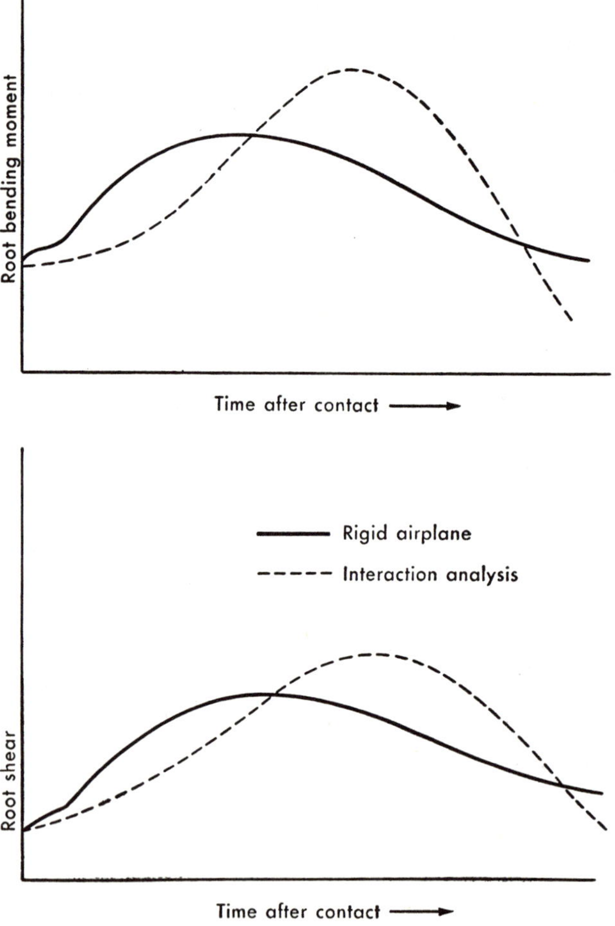

FIG. 7–4. Effects of interaction on dynamic response.

those for a rigid body, depending largely on the ratio of the duration of the impact force to the natural period of the structure.

An illustration of the effects of interaction is shown in Fig. 7–4, where bending moments and shears for a rigid structure subjected to a rigid-body forcing function are compared with an interaction solution (Reference 18).

7–4. Gust Response and Buffeting. The gust response problem is similar in many respects to the landing impact problem discussed in the preceding article. The main differences reside in the fact (1) that no mechanical system (i.e., landing gear) exists to transmit the applied loads to the structure, but instead the applied loads are entirely aerodynamic, and (2) that for the landing problem the loads are applied at a known number of discrete points on the structure, while for the gust problem the load is distributed over the structure according to some time and space variation. A feature of similarity in the two problems is the possibility that the flexibility of the structure may cause significant modifications in the applied loads. In the present case this means changes in the aerodynamic forces resulting from structural distortions, in the manner discussed in Chapters 5 and 6.

The usual type of gust analysis used in design is based upon the concept of a sharp-edged gust of velocity w, with the additional assumptions that the airplane is instantaneously immersed in the gust and that the aerodynamic forces are instantaneously adjusted to steady-state values. Thus, there is a change in airplane lift (per unit span) given by the expression

$$\Delta L = \rho b V^2 \left(\frac{dC_L}{d\alpha}\right)\left(\frac{w}{V}\right) \tag{7–42}$$

where b is the wing semichord. If we neglect pitching of the airplane, or any unsteady aerodynamic effects, the equation of motion of the airplane can be written very simply as

$$m\ddot{z} = \rho b V^2 \left(\frac{dC_L}{d\alpha}\right)\left(\frac{w}{V} + \frac{\dot{z}}{V}\right) \tag{7–43}$$

from which the airplane normal acceleration can be found as

$$\ddot{z} = -\frac{Vw}{2\lambda b}\, e^{-(s/2\lambda)} \tag{7–44}$$

where

$$\lambda = \frac{2b^2 m}{\rho}\left(\frac{dC_L}{d\alpha}\right) \tag{7–45a}$$

$$s = \frac{V}{b}\, t \tag{7–45b}$$

From this result an incremental load factor Δn can be written as

$$\Delta n = \frac{\rho b V w}{mg}\left(\frac{dC_L}{d\alpha}\right) \tag{7-46}$$

A more detailed study of a gust load formula suitable for design purposes may be found in Reference 20.

The assumptions of this simple analysis are not satisfied, of course, in the actual case. Thus, the velocity w must be regarded as that of an equivalent sharp-edged gust that will produce the same normal acceleration of the airplane under the idealized conditions as does the real gust under actual conditions. The relationship between the actual gust velocity and that of the equivalent gust depends in large measure upon the

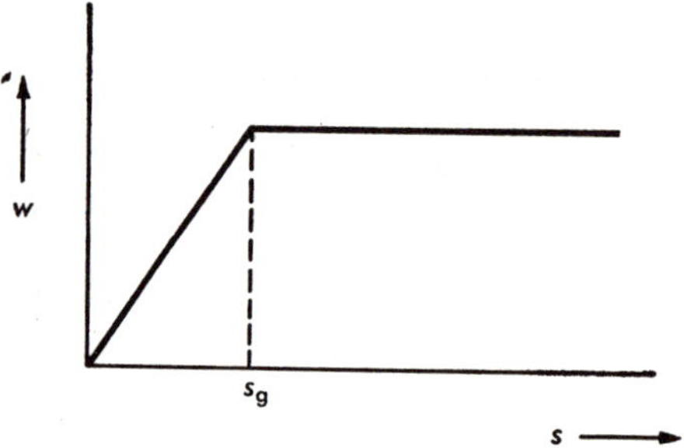

FIG. 7–5. Gust with linear build-up.

characteristics of the particular airplane, since the maximum lift does not occur until after the airplane has penetrated some distance into the gust and has attained some velocity in the direction of the gust. In order to take some account of this effect in determining loads from the preceding simple, sharp-edged gust formulas, an "alleviation factor" is introduced. The alleviation factor depends upon the wing loading, the wing geometry, the airplane-lift curve slope, the airplane pitching motion, etc.

Considering a simple gust with linear build-up, as shown in Fig. 7–5, the alleviation factor can be determined from steady-flow theory as (Reference 19)

$$AF = \frac{1 - e^{-(s_g/2\lambda)}}{s_g/2\lambda} \tag{7-47}$$

where s_g denotes the value of the dimensionless variable s at which the gust velocity becomes constant. A plot of the alleviation factor versus s_g for a particular value of λ is shown in Fig. 7–6, while a plot of normal acceleration is shown in Fig. 7–7.

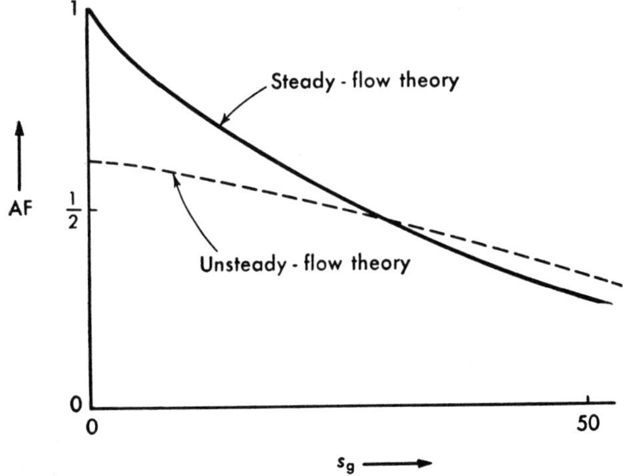

FIG. 7–6. Alleviation factor for gust with linear build-up (rigid wing).

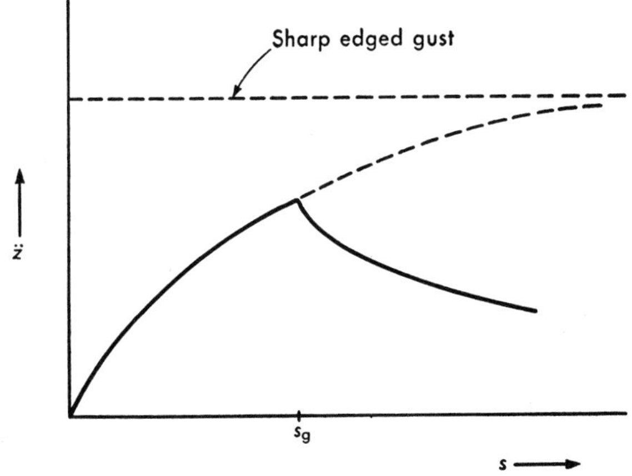

FIG. 7–7. Normal acceleration for gust with linear build-up (rigid wing).

It is clear, however, that the motion of a wing penetrating a gust must be greatly influenced by the ensuing motion of the wing, and hence a more realistic analysis would include unsteady-flow effects. The alleviation factor for the gust described above, when based on unsteady-flow theory, is shown also in Fig. 7–6. It seems worthwhile, then, to describe briefly the nature of the unsteady aerodynamic forces which act on the wing.*

* See also Art. 5–4.

These forces arise from the gust itself and from the ensuing motion of the airplane. One manner in which these forces may be studied involves considering the lift and moment in two parts: one part includes all lift forces (exclusive of aerodynamic inertia effects) and thus refers to circulation, while the other part includes only inertia effects. The various forces and moments on a wing can be resolved as shown in Fig. 7–8, where L_g is the lift force produced by the gust and L_1, L_2, L_3, and M_a arise as a consequence of the wing motion.

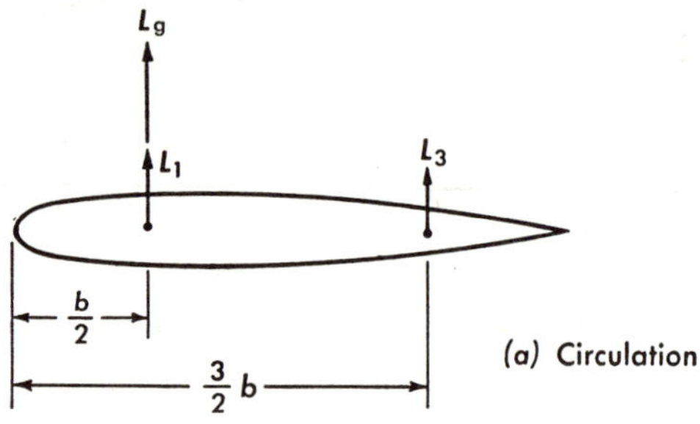

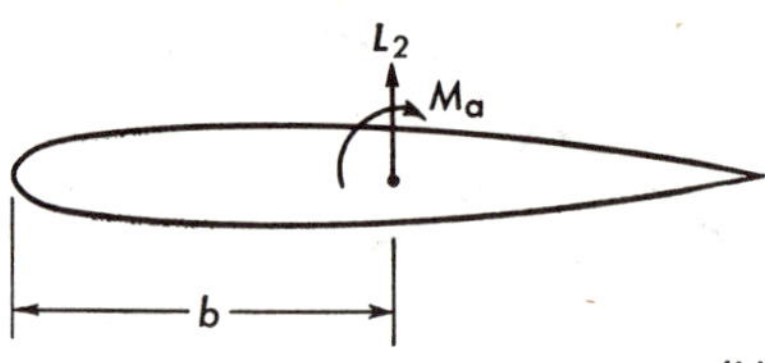

FIG. 7–8. Forces and moments on an airplane wing during gust loading.

The lift force per unit span produced by a gust of vertical velocity can be written in the following form, if the time of penetration of the gust is denoted by τ:

$$L_g = 2\pi\rho bV \int_0^t \frac{\partial w}{\partial t} \Psi(t - \tau)\, d\tau \tag{7–48}$$

where V is the forward velocity of the wing, and Ψ is the Küssner function defined in Chapter 5. The other lift forces due to circulation result from the motion of the wing and can be written as

$$L_1 = 2\pi\rho bV \int_0^t [(\tfrac{1}{2} - a)b\ddot{\alpha} + V\dot{\alpha} + \ddot{h}]\varphi(t - \tau)\, d\tau \tag{7–49}$$

$$L_3 = \pi\rho b^2 V\dot{\alpha} \tag{7–50}$$

where h is the vertical deflection of the elastic axis (located by a) and is positive downward, α is the rotation of the airfoil section and is positive when stalling, and φ is the Wagner function, also defined in Chapter 5. The Küssner function, Ψ, gives the growth in lift as an airfoil penetrates a sharp-edged vertical gust, while the Wagner function, φ, gives the lift due to the growth of circulation on the airfoil section following a sudden change in angle of attack (see Fig. 5–3). A collection of unsteady-flow functions is given in Reference 21.

The inertia force and moment can be written in a similar manner as

$$L_2 = \pi\rho b^2(\ddot{h} - ab\ddot{\alpha}) \tag{7–51}$$

$$M_a = -\frac{\pi\rho b^4}{8}\ddot{\alpha} \tag{7–52}$$

The total lift and moment (about the elastic axis) per unit span are then given as

$$L_T = L_1 + L_2 + L_3 + L_g \tag{7–53}$$

$$M_T = b(\tfrac{1}{2} + a)(L_1 + L_g) + abL_2 - (\tfrac{1}{2} - a)bL_3 + M_a \tag{7–54}$$

Previously, the discussion was directed toward the gust response of a rigid wing, although the foregoing unsteady-flow analysis takes into account both bending and torsional deformations of the wing. The problem of analyzing the dynamic response of a flexible wing to a gust load now reduces to (1) defining the differential displacement equations (which may be taken, for example, to be the equations of simple beam bending and twist), and (2) of considering Eqs. 7–53 and 7–54 to represent the applied transverse and torque loadings. Thus, if we consider the wing to be represented by a simple free-free beam, we have the equation

$$\frac{\partial^2}{\partial y^2}\left(EI\frac{\partial^2 h}{\partial y^2}\right) = -m\ddot{h} + L_T \tag{7–55}$$

The introduction of wing flexibility brings out two new aspects of the problem. First, the inclusion of wing flexibility may show that higher stresses exist than would be indicated by a rigid-body analysis; and second, wing flexibility may introduce some error when an airplane is used as an instrument for measuring gust intensity. Studies involving flexible wing treatments have been given by a number of authors (References 4, 6, 22–24), employing both modal and other representations of the deformed structure. Reference 23 analyzes the structural response to gusts of an airplane having freedom of vertical motion and wing bending flexibility. A convenient numerical solution of the response equations is given for the case of a discrete gust encounter, and an exact solution is given for the simpler case of encounter of a continuous sinusoidal gust.

The inadequacy of analyses based on discrete gust encounter has long been suspected, and it has been replaced recently by a concept of random, continuous atmospheric disturbances (Reference 25). Thus, it has come to be recognized that there are many dynamic stress problems for which the knowledge of the physical phenomena is not precise enough to justify exact predictions with respect to the results of an individual observation. On one hand, the loading under such conditions may never be known exactly, and on the other hand, the physical system may be so complex that detailed calculation of dynamic response is difficult and uncertain. This is certainly true of the modal methods and particularly so if trend studies are desired. In such cases the physical data can be regarded as a set of statistical data relating to random processes. It is clear, then, that any truly logical statements regarding the forcing functions must be statistical, and the objective of analysis must be to make valid inferences about the responses by statistical means.

A prime example of such a random type of disturbance is buffeting. In its most common and simple form, buffeting occurs as a response of the tailplane to eddying flow from the wing wake. In this instance, buffeting is a forced vibration caused by fluctuations in the wing wake, and the tail must have its own natural frequencies far removed from that of the wake in order to avoid resonance. The wake fluctuations are often described in terms of the Kármân vortex street, and as such, the frequency of the pulsating forces is given by

$$\omega = \frac{kv}{s} \tag{7–56}$$

where k is a constant and s is a distance. Because the frequency spectrum in the wake covers such a wide range, and because of other random factors, it does not seem possible to give analyses other than on a statistical basis. The analyses of gust and buffeting problems, as well as the landing impact problem, by statistical methods are receiving much attention (References 14, 23, 26–31).

7–5. Additional Remarks. It should not be inferred that landing impact, gust loading, and buffeting are the only impulsive loads to which aircraft are subjected. Also of importance are the dynamic responses arising in such problems as the sloshing of fuel in tanks (Reference 28), blast waves (Reference 32), catapult take-off, and rapid maneuvers.

While it is essential to realize that each of these problems is wholly different from the others in detail, it is equally essential to be aware of the fact that all are problems of impulsive loading and hence may be treated by various of the general methods mentioned in this chapter. In connection with this, it is perhaps advisable to make some comments relative to the use of the modal method in problems of impulsive loading.

An elastic structure subjected to impulsive forces responds in a dynamical manner. The problem may be studied from several points of view, one of which is the superposition of natural vibrations (modal method) and another of which is the difference-differential equation method in which equilibrium conditions are satisfied at certain points in the structure (Reference 5). In the former method, to be rigorous, the superposition of an infinite number of natural vibration modes is required; but, of course, in practical computations only a small number can be actually taken.

An analysis of the general theory of unrestrained elastic structures by the normal mode procedure leads to equations in the form of infinite series to express the deformations, moments, and shears; each term of the series represents a mode of vibration. If we are to use only the first few terms of such series, we must naturally be concerned with the question of convergence. For problems not involving impulsive loading, few difficulties of this type arise; however, when dealing with problems of dynamic response, such difficulties are often encountered. For example, in the case of a simply supported rectangular beam subjected to a unit impulse, that is,

$$\int F(t)\, dt = 1 \tag{7–57}$$

treated by the modal method, it is found (Reference 1) that the contributions of the second, third, and fourth modes to the moment are of the same order of magnitude as that of the first mode. Even worse, the contributions of the second, third, and fourth modes to the shear are three, five, and seven times as much, respectively, as that of the first mode. In the case of a unit-step load (Fig. 7–1), the situation is not quite so bad, the higher modes having an appreciable effect only on the shear. Limitations of this kind appear in all elastic structures.

Another question of importance is the distribution of loads on the structure. For loads distributed over all or most of the structure, the modal method may give quite acceptable results. For only a few concentrated loads, however, it would be expected that a great many modes will be excited, and hence, when considering only a few modes, accurate results are not likely to be obtained. If the duration of the loading is small compared with the period of the fundamental mode, all higher modes whose periods are not small compared with the duration of loading must be considered. This helps to explain why the unit impulse loading is so much more critical than the unit-step function for the beam, as described above. The time rate of loading also has a very marked effect. Finally, it may be noted that the modal method may not yield good results in those cases where there is resonance between slightly coupled components of the system because of the transfer of energy from one component to another. Some additional comparisons of modal and other methods are given in Reference 16.

In summing up this discussion of modal methods it is emphasized that such methods may be used with good results if the modes used (both number and shape) are carefully selected, and if the loading condition is not too severe. In such cases other methods of analysis may prove superior.

PROBLEMS

7–1. Prove that for a load applied suddenly and then held constant, as shown in Fig. 7–1, the value of the dynamic load factor is 2.0.

7–2. For the applied force shown in Fig. 7–9, determine the maximum value of the dynamic load factor.

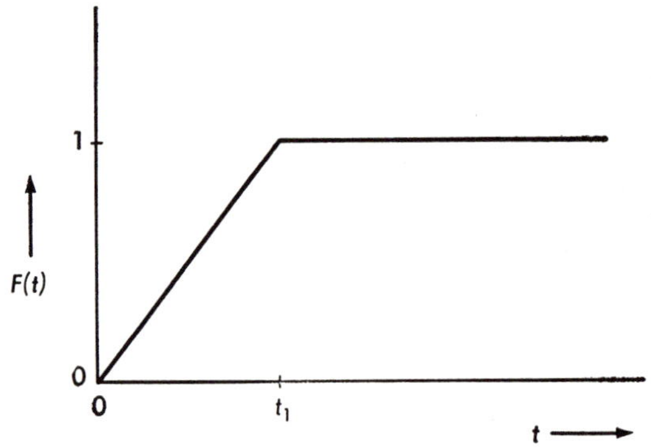

Fig. 7–9. Problem 2.

7–3. The differential equation for longitudinal vibration of a rod is of the form

$$AEg\,\frac{\partial^2 u(x, t)}{\partial x^2} = \frac{w}{g}\,A\,\frac{\partial^2 u(x, t)}{\partial t^2} - gF(x, t)$$

Derive an expression for the dynamic response of the rod under any boundary conditions, in terms of the normal modes, the participation factor, and the dynamic load factor.

7–4. Consider a simply supported beam subjected to a unit-step function excitation which is applied uniformly along the span of the beam. Determine the participation factor for the first three modes.

7–5. Derive Eq. 7–44.

7–6. Compare Eqs. 7–48 to 7–54 with Eqs. 5–31, 5–33 to 5–35, and 5–38. Discuss the relationship between the two sets of equations.

REFERENCES

1. Biot, M. A., and Bisplinghoff, R. L. "Dynamic Loads on Aircraft Structures During Landing," *NACA Wartime Report W-92* (ARR 4H10) (1944).

2. Timoshenko, S. P. *Vibration Problems in Engineering*. 3d ed., New York: D. Van Nostrand Co., Inc., 1955. Ch. V, pp. 329–331.
3. Frankland, J. M. "Effects of Impact on Simple Elastic Structures," *Proc. Soc. Exp. Stress Analysis*, **6** (1948): 7–27.
4. Bisplinghoff, R. L., Isakson, G., and Pian, T. H. H. "Methods in Transient Stress Analysis," *Jour. Aero. Sciences*, **17** (May, 1950): 259–270.
5. Houbolt, J. C. "A Recurrence Matrix Solution for the Dynamic Response of Elastic Aircraft," *Jour. Aero. Sciences*, **17** (September, 1950): 540–550.
6. Jenkins, E. S., and Pancu, C. D. P. "Dynamic Loads on Airplane Structures," *SAE Quart. Transactions*, **3** (1949): 391–409.
7. Milwitzky, B., and Cook, F. E. "Analysis of Landing-Gear Behavior," *NACA Tech. Note* 2755 (1952).
8. Flügge, W. "Landing-Gear Impact," *NACA Tech. Note* 2743 (1952).
9. Stowell, E. Z., Houbolt, J. C., and Batdorf, S. B. "An Evaluation of Some Approximate Methods of Computing Landing Stresses in Aircraft," *NACA Tech. Note* 1584 (1948).
10. Goland, M., Luke, Y. L., and Kahn, E. A. "Prediction of Dynamic Landing Loads," *Air Force Technical Report* 5815 (1949).
11. Pian, T. H. H., and Flomenhoft, H. I. "Analytical and Experimental Studies on Dynamic Loads in Airplane Structures During Landing," *Jour. Aero. Sciences*, **17** (December, 1950): 765–774.
12. Zahorski, A. H. "Remarks on Dynamic Loads in Landing," *Jour. Aero. Sciences*, **19** (April, 1952): 258–264.
13. Rosenbaum, R. "Outline of an Acceptable Method of Determining Dynamic Landing Loads," *CAA Airframe and Equipment Engineering Report No.* 52 (1954).
14. Fung, Y. C. "The Analysis of Dynamic Stresses in Aircraft Structures During Landing as Nonstationary Random Processes," *Jour. App. Mechanics*, **22** (December, 1955): 449–457.
15. Kroll, W. D., and Levy, S. "A Step-by-Step Method of Determining the Dynamic Response of Aircraft in Landing," *National Bureau of Standards Rept.* 6.4/1–181, *PR* 6.
16. Ramberg, W. "Transient Vibration in an Airplane Wing Obtained by Several Methods," National Bureau of Standards, *Jour. of Research*, **42** (May, 1949): 437–447.
17. McPherson, A. E., Evans, J., and Levy, S. "Influence of Wing Flexibility on Force-Time Relation in Shock Strut Following Vertical Landing Impact," *NACA Tech. Note* 1995 (1949).
18. Cook, F. E., and Milwitzky, B. "Effect of Interaction on Landing Gear Behavior and Dynamic Loads in a Flexible Airplane Structure," *NACA Rept.* 1278 (1956).
19. Bisplinghoff, R. L., Isakson, G., and O'Brien, T. F. "Gust Loads on Rigid Airplanes with Pitching Neglected," *Jour. Aero. Sciences*, **18** (January, 1951): 33–42.
20. Pratt, K. G., and Walker, W. G. "A Revised Gust Load Formula and a Re-Evaluation of V-G Data Taken on Civil Transport Airplanes from 1933 to 1950," *NACA Rept.* 1206 (1954).
21. Drischler, J. A. "Calculation and Compilation of the Unsteady Lift Functions for a Rigid Wing Subjected to Sinusoidal Gusts and Sinusoidal Sinking Oscillations," *NACA Tech. Note* 3748 (1956).
22. Houbolt, J. C. "A Recurrence Matrix Solution for the Dynamic Response of Aircraft in Gusts," *NACA Tech. Note* 2060 (1950).
23. Houbolt, J. C., and Kordes, E. E. "Gust Response Analysis of an Airplane, Including Wing Bending Flexibility," *NACA Rept.* 1181 (1954).
24. Goland, M., Luke, Y. L., and Kahn, E. A. "Prediction of Wing Loads Due to Gusts, Including Aero-Elastic Effects," *Air Force Tech. Rept.* 5706 (1947).
25. Donely, P. "Summary of Information Relating to Gust Loads on Airplanes," *NACA Rept.* 997 (1950).

26. Press, H., and Houbolt, J. C. "Some Applications of Generalized Harmonic Analysis to Gust Loads on Airplanes," *Jour. Aero. Sciences*, **22** (January, 1955): 17–26.
27. Diederich, F. W. "The Dynamic Response of a Large Airplane to Continuous Random Atmospheric Disturbances," *Jour. Aero. Sciences*, **23** (October, 1956): 917–930.
28. Luskin, H., and Lapin, E. "An Analytical Approach to the Fuel Sloshing and Buffeting Problems of Aircraft," *Jour. Aero. Sciences*, **19** (April, 1952): 217–228.
29. Liepmann, H. W. "On the Application of Statistical Concepts to the Buffeting Problem," *Jour. Aero. Sciences*, **19** (December, 1952): 793–800.
30. Fung, Y. C. "Statistical Aspects of Dynamic Loads," *Jour. Aero. Sciences*, **20** (May, 1953): 317–330.
31. Ribner, H. S. "Spectral Theory of Buffeting and Gust Response: Unification and Extension," *Jour. Aero. Sciences*, **23** (December, 1956): 1075–1077.
32. Bisplinghoff, R. L., and Witmar, E. A. "Blast Loading of Aircraft Structures," *Sixth Anglo-American Aero. Conf.* (1957).

CHAPTER 8

INTRODUCTION TO FLIGHT STABILITY

8–1. Introduction. An airplane or missile in flight through the earth's atmosphere is acted upon by inertia forces resulting from changes in the flight path, by the force of gravity, by the forces of its propulsive unit, and by the aerodynamic forces and moments arising from its motion relative to the atmosphere.

In order for the aircraft to maintain a steady, straight flight path, the various forces acting on it must be in static equilibrium, while if the flight path is unsteady or curved in any way, the forces acting on it must be in dynamic equilibrium. Generally speaking, the flight path is controlled and determined by the pilot, whether human or automatic; however, disturbances or perturbations to the flight path often occur, and for proper flying qualities of the aircraft it is necessary that the ensuing motion be stable. Such disturbances may arise from atmospheric gusts or turbulences or from abrupt manuevers; in the former case, a stable motion is required in order that the pilot may make a satisfactory recovery, while in the latter case, a stable motion is required to enable the pilot to retain control of the aircraft during the maneuver and thus effect the subsequent recovery.

In a broad sense, the subject of flight stability may be divided into two groups, *static* stability and *dynamic* stability (see Art. 4–2). An aircraft is considered to be statically stable if, following a disturbance, forces and moments are set up which tend to restore the aircraft to its original flight path. The aircraft is considered to be dynamically stable if the oscillatory flight path following a disturbance is eventually damped out. It follows that dynamic stability implies static stability, but the mere possession of static stability does not ensure dynamic stability.

The rigid-body motion of an airplane may also be classified in two groups: longitudinal and lateral motions. Longitudinal motions are those which occur in the plane of symmetry of the airplane, while lateral motions are those which displace the plane of symmetry, i.e., rolling and yawing motions. The rolling and yawing motions cannot be separated from one another since a velocity in roll causes a change in both the rolling and yawing moments acting on the airplane. Similarly, a velocity in yaw causes a change in both rolling and yawing moments, and finally, a sideslip velocity results in both rolling and yawing moments. The fact that these two types of motion cannot be separated accounts for the relative complexity of lateral stability analyses when compared with the case of longitudinal stability.

Longitudinal dynamic instability manifests itself as a divergent oscillation in pitching motion, while lateral dynamic instability may manifest itself in two ways, following a disturbance. In the first, the motion is an increasing spiral motion, called *spiral instability*, and in the second, the motion consists of increasing directional (yawing) oscillations, accompanied by rolling, called *Dutch roll.*

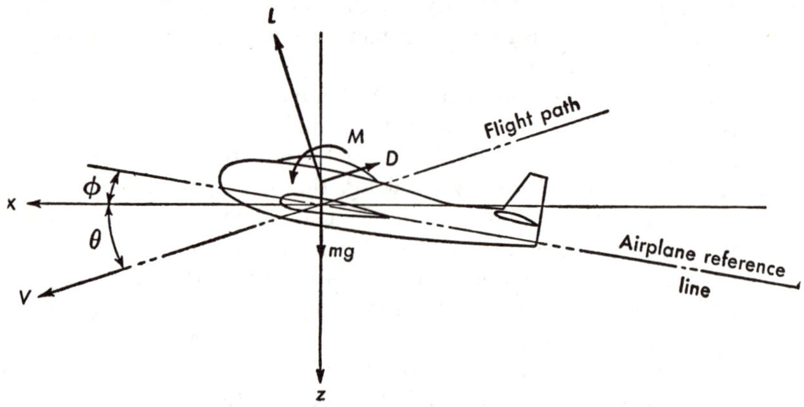

FIG. 8–1. Forces and moments on an airplane in longitudinal motion.

As a very simple introduction to the type of analysis that is required in stability studies, let us consider the longitudinal motion of a rigid airplane in power-off flight, assuming that the motion consists of pitching and displacement in the plane of symmetry. Referring to the forces, moments, and orientation of the reference axes shown in Fig. 8–1, we may apply Newton's law to obtain the equations of motion

$$m \frac{dV}{dt} = -D + mg \sin \theta \tag{8–1}$$

$$mV \frac{d\theta}{dt} = -L + mg \cos \theta \tag{8–2}$$

$$I \frac{d^2\phi}{dt^2} = -M \tag{8–3}$$

The lift, drag, and moment are denoted by L, D, and M, respectively; θ and ϕ denote flight path angle and the airplane reference angle, respectively; while m is the airplane mass, I is its moment of inertia, and V is its velocity.

We concern ourselves only with small disturbances from the normal flight path, so that θ and ϕ are small angles and $V = V_0 + v$, where $v \ll V_0$. For simplicity, assume that the drag is negligible compared with the other forces. The equations of motion then become

$$\dot{v} = g\theta \tag{8-4}$$

$$\dot{\theta} = -\frac{L}{mv} + \frac{g}{v} \tag{8-5}$$

$$\ddot{\phi} = -\frac{M}{I} \tag{8-6}$$

In level, unaccelerated flight the lift is equal to the airplane weight, the pitching moment is zero, the reference angle ϕ is zero, and the angle of attack is a constant α_0. Thus, we may assume that in the disturbed flight

$$L = mg + \Delta L \tag{8-7}$$

$$M = \Delta M \tag{8-8}$$

$$\alpha = \alpha_0 + \Delta\alpha \tag{8-9}$$

$$\Delta\alpha = \theta + \phi \tag{8-10}$$

Because the lift depends upon the square of the velocity and the angle of attack, we may write

$$L = mg\left(\frac{V}{V_0}\right)^2\left(\frac{\alpha}{\alpha_0}\right) \tag{8-11}$$

Upon introducing the unsteady values for V and α, we find, omitting higher order terms,

$$L = mg\left(1 + 2\frac{v}{V_0} + \frac{\theta + \phi}{\alpha_0}\right) \tag{8-12}$$

The aerodynamic pitching moment M is composed of two parts: the aerodynamic moment of the wing and the moment provided by the aerodynamic force acting on the tailplane. The latter also is composed of two parts: one due to change in angle of attack, and one due to rotation of the airplane. The part of the total moment contributed by the wing and by that part of the tail moment resulting from changes in angle of attack can be written as

$$\Delta M_1 = mgck_s\,\Delta\alpha \tag{8-13}$$

where c is the mean wing chord and k_s is a numerical constant which governs the magnitude of the restoring force and hence is important in determining the static stability. The contribution to the total moment provided by the rotation of the airplane will depend essentially upon the "tail length" l, i.e., the distance from the airplane c.g. to the center of pressure of the tail and the velocity of rotation $\dot{\phi}$. The rotation thus produces a force on the tail proportional to its effective angle of attack, which may be written as $l\dot{\phi}/V_0$. The contribution to the total moment may then be expressed in the form

$$\Delta M_2 = mgk_t l^2 \frac{\dot{\phi}}{V_0} \tag{8–14}$$

where k_t is also a numerical constant.

Eqs. 8–12, 8–13, and 8–14 may be substituted into Eqs. 8–4, 8–5, and 8–6, and upon eliminating v we find

$$\frac{d^2\theta}{d\tau^2} + \frac{1}{\alpha_0}\left(\frac{d\theta}{d\tau} + \frac{d\phi}{d\tau}\right) + 2\theta = 0 \tag{8–15}$$

$$\frac{d^2\phi}{d\tau^2} + \delta\frac{d\phi}{d\tau} + \sigma(\phi + \theta) = 0 \tag{8–16}$$

where

$$\tau = \frac{g}{V_0}t \tag{8–17}$$

$$\delta = k_t\frac{ml^2}{I} \tag{8–18}$$

$$\sigma = k_s\frac{mcV_0^2}{gI} \tag{8–19}$$

Eqs. 8–15 and 8–16 represent a convenient form of the equations of motion for further analysis.

The parameter δ indicates the damping effect of the horizontal tail, while σ indicates the static stability characteristics of the airplane. If $\sigma > 0$, then a rotation of the airplane through an angle ϕ will produce a restoring moment so that the airplane is statically stable. If $\sigma < 0$, then a rotation of the airplane through an angle ϕ will produce a moment which tends to increase ϕ so that the airplane is statically unstable.

We now propose to study the dynamic stability characteristics of the airplane by seeking solutions of Eqs. 8–15 and 8–16. As trial solutions we may choose

$$\theta = Ae^{\lambda\tau}, \quad \phi = Be^{\lambda\tau} \tag{8–20}$$

which leads, by expansion of the resulting determinantal equation, to the frequency equation

$$\lambda^4 + \left(\frac{1}{\alpha_0} + \delta\right)\lambda^3 + \left(2 + \sigma + \frac{\delta}{\alpha_0}\right)\lambda^2 + 2\delta\lambda + 2\sigma = 0 \tag{8–21}$$

The stability characteristics, as provided by this quartic equation, may be investigated by the methods outlined in Chapter 4. It was found there that for the general quartic equation

$$\lambda^4 + a_3\lambda^3 + a_2\lambda^2 + a_1\lambda + a_0 = 0 \tag{8-22}$$

the conditions for stability are given by

$$a_0, a_1, a_2, a_3 > 0 \tag{8-23a}$$

$$a_1^2 - a_3(a_2a_1 - a_3a_0) < 0 \tag{8-23b}$$

In terms now of our problem, Eq. 8–21, we note at once that α_0 and δ are positive by definition, so that the conditions of positive coefficients will be satisfied if $\sigma > 0$. This means, then, that the aircraft will be dynamically stable only if it is also statically stable. This was noted earlier by intuitive reasoning. Now, the condition that the discriminant must be negative leads to the inequality

$$\delta^2 + \frac{2\delta\alpha_0}{1 + \delta\alpha_0} > \sigma \tag{8-24}$$

For a given value of α_0, we may plot this relation in a $\sigma - \delta$ diagram, as shown in Fig. 8–2. The regions of stability are as indicated, and the effect of considering the drag is also shown.

A very interesting type of motion can be studied if we consider an airplane which possesses large static stability so that σ is very large. The

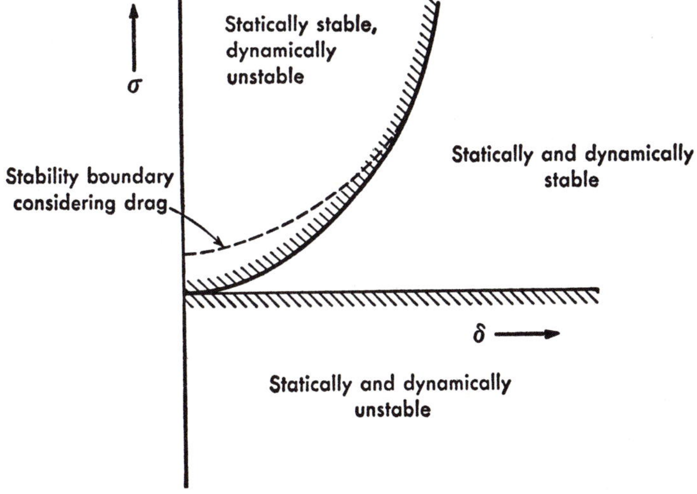

FIG. 8–2. Longitudinal stability chart.

frequency equation (Eq. 8–21) then reduces to

$$\lambda^2 + 2 = 0 \tag{8–25}$$

and the equations of motion become

$$\begin{aligned} \frac{d^2\theta}{d\tau^2} + \frac{1}{\alpha_0}\frac{d}{d\tau}(\theta + \phi) + 2\theta &= 0 \\ \frac{d^2\phi}{d\tau^2} + \sigma(\theta + \phi) &= 0 \end{aligned} \tag{8–26}$$

The solutions of these two equations may be expressed as

$$\begin{aligned} \phi &= -\theta \\ \theta &= C_1 \sin \sqrt{2}\,\tau + C_2 \cos \sqrt{2}\,\tau \end{aligned} \tag{8–27}$$

where C_1 and C_2 are arbitrary constants to be found from the initial conditions. The period of this oscillatory motion is

$$T_p = 2\pi \frac{V_0}{g\sqrt{2}} \tag{8–28}$$

This motion characterizes an airplane with large restoring moment and small moment of inertia, and the motion is such that, since

$$\Delta\alpha = \theta + \phi \approx 0$$

the angle of attack remains essentially constant. This type of motion is called *phugoid* motion.

Because the frequency equation (Eq. 8–21) is a quartic, there must be two roots in addition to $\lambda = \pm i\sqrt{2}$ as given by Eq. 8–25. This means that λ^4 and $\sigma\lambda^2$ are of the same order, so that the simplified frequency equation is more correctly expressed as

$$\lambda^4 + \sigma\lambda^2 + 2\sigma = 0 \tag{8–29}$$

Two roots of this equation are given approximately by Eq. 8–25, while two additional roots are approximately given by

$$\lambda^2 + \sigma = 0 \tag{8–30}$$

The corresponding equations of motion become

$$\begin{aligned} \theta &= 0 \\ \frac{d^2\phi}{d\tau^2} + \sigma\phi &= 0 \end{aligned} \tag{8–31}$$

and the solution is then

$$\phi = C_1' \sin \sqrt{\sigma}\,\tau + C_2' \cos \sqrt{\sigma}\,\tau \tag{8–32}$$

The motion is an oscillatory, high-frequency one about the airplane c.g. and has a period given by

$$T_s = 2\pi \frac{V_0}{g\sqrt{\sigma}} = 2\pi \sqrt{\frac{I}{mgck_s}} \tag{8-33}$$

and because the restoring moment is large and the moment of inertia is large, this period is quite small. This motion is often referred to as the

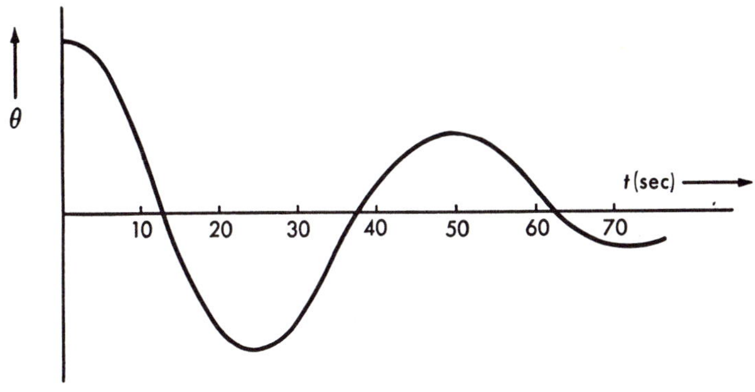

(a) Phugoid mode

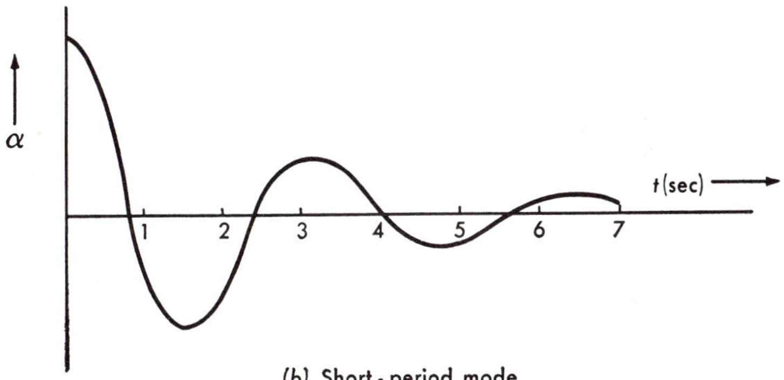

(b) Short-period mode

FIG. 8–3. Modes of longitudinal motion.

short-period longitudinal motion of the airplane. Changes in forward velocity during this motion are negligible.

Typical phugoid and short-period modes of longitudinal motion are shown in Fig. 8–3.

8–2. Equations of Longitudinal Motion. For writing the equations for small, disturbed motions of an airplane in power-off symmetrical flight, two different coordinate axes systems may be employed, both of which, however, are fixed in the airplane and move with it. One of these systems, the *body axes* system, involves a set of axes fixed in the airplane with the x-axis coincident with the airplane reference line (Fig. 8–4a). The other, the *wind axes* system, considers that the x-axis always points into the

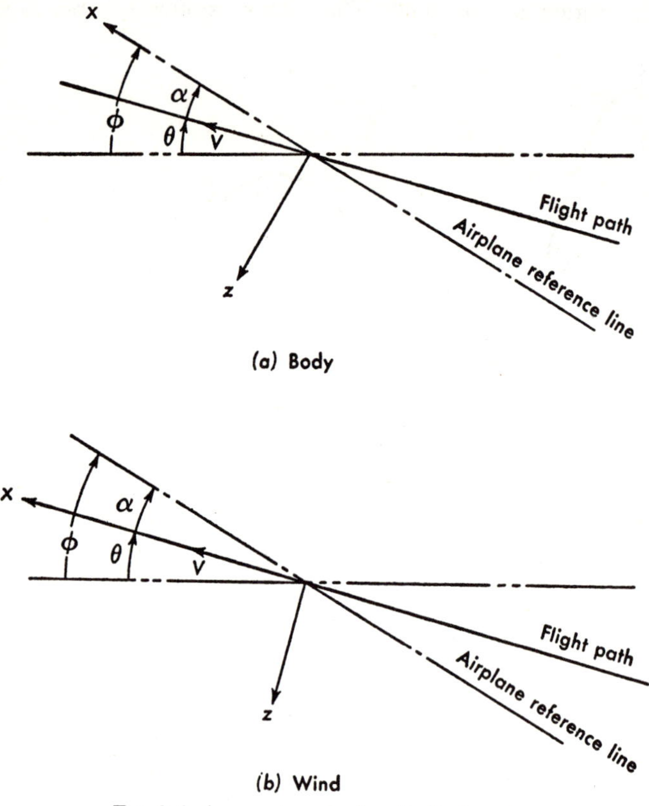

FIG. 8–4. Axes systems for longitudinal motion.

direction of the relative wind (Fig. 8–4b), and hence there is a slight rotation of the axes within the airplane as it moves along its flight path.

Considering now the airplane axes as the wind axes, the symmetric equations of motion are

$$\begin{aligned} m\dot{V} &= X \\ -mV\dot{\theta} &= Z \\ I_y\dot{q} &= M \end{aligned} \tag{8-34}$$

where X and Z are components of the resultant force and $q \equiv \dot{\phi}$. We shall consider that the variables of the problem are the disturbances from straight, steady flight, so that the resultant force components and the moment can be expressed as

$$\begin{aligned} X &= X_0 + \frac{\partial X}{\partial V} dV + \frac{\partial X}{\partial \alpha} d\alpha + \frac{\partial X}{\partial \phi} d\phi + \frac{\partial X}{\partial \dot{V}} d\dot{V} \\ &\quad + \frac{\partial X}{\partial \dot{\alpha}} d\dot{\alpha} + \frac{\partial X}{\partial \dot{\phi}} d\dot{\phi} \\ Z &= Z_0 + \frac{\partial Z}{\partial V} dV + \frac{\partial Z}{\partial \alpha} d\alpha + \frac{\partial Z}{\partial \phi} d\phi + \frac{\partial Z}{\partial \dot{V}} d\dot{V} \\ &\quad + \frac{\partial Z}{\partial \dot{\alpha}} d\dot{\alpha} + \frac{\partial Z}{\partial \dot{\phi}} d\dot{\phi} \\ M &= M_0 + \frac{\partial M}{\partial V} dV + \frac{\partial M}{\partial \alpha} d\alpha + \frac{\partial M}{\partial \phi} d\phi + \frac{\partial M}{\partial \dot{V}} d\dot{V} \\ &\quad + \frac{\partial M}{\partial \dot{\alpha}} d\dot{\alpha} + \frac{\partial M}{\partial \dot{\phi}} d\dot{\phi} \end{aligned} \tag{8-35}$$

where X_0, Z_0, and M_0 represent the values for straight, steady flight. Now, the components of the resultant force will depend almost entirely upon the changes in V, α, and ϕ, and only slightly upon the time derivatives of these variables. Similarly, the moment will be only very slightly influenced by changes in ϕ and $\dot{V}$. Denoting the partial derivatives by subscripts (e.g., $\partial X/\partial\alpha \equiv X_\alpha$), for convenience, the equations (Eq. 8–35) reduce to

$$\begin{aligned} X &= X_0 + X_v\, dV + X_\alpha\, d\alpha + X_\phi\, d\phi \\ Z &= Z_0 + Z_v\, dV + Z_\alpha\, d\alpha + Z_\phi\, d\phi \\ M &= M_0 + M_v\, dV + M_\alpha\, d\alpha + M_{\dot{\alpha}}\, d\dot{\alpha} + M_{\dot{\phi}}\, d\dot{\phi} \end{aligned} \tag{8-36}$$

Now, when the airplane is slightly disturbed from steady flight, we may represent the various quantities as

$$\begin{aligned} V &= V_0 + V', \quad (V' \ll V_0) \\ \alpha &= \alpha_0 + \alpha', \quad (\alpha' \ll \alpha_0) \\ \phi &= \phi_0 + \phi', \quad (\phi' \ll \phi_0) \end{aligned} \tag{8-37}$$

etc.

Changes in the variables may then be represented by such relations as

$$\Delta V = V - V_0 = V' \tag{8-38}$$

Dropping the primes on the small quantities, for convenience, and also denoting V' by v, the symmetric equations of motion (Eq. 8–34) can finally be written in the abbreviated form

$$\begin{aligned} m\dot{v} &= vX_v + \alpha X_\alpha + \phi X_\phi \\ -mv\dot{\theta} &= vZ_v + \alpha Z_\alpha + \phi Z_\phi \\ mk_y^2\dot{q} &= vM_v + \alpha M_\alpha + \dot{\alpha}M_{\dot{\alpha}} + \dot{\phi}M_{\dot{\phi}} \end{aligned} \tag{8–39}$$

where k_y is the radius of gyration about the y-axis. The various partial derivatives are known collectively as *stability derivatives.*

To evaluate the forms of the various stability derivatives, we require some knowledge of the forces acting on the airplane. For power-off flight the only significant forces are the lift, drag, and weight, and since we use wind axes, it is particularly easy to obtain the various derivatives. Thus, we have

$$X = -D - W(\phi - \alpha) \tag{8–40a}$$

or

$$X = -\tfrac{1}{2}\rho SV^2C_D - W(\phi - \alpha) \tag{8–40b}$$

The quantities V, ϕ, and α here denote the values for disturbed flight and hence must be replaced by Eq. 8–37. Omitting higher order terms, the stability derivatives are found to be

$$\begin{aligned} X_v &= \frac{\partial X}{\partial v} = -\rho SV_0C_D \\ X_\alpha &= \frac{\partial X}{\partial \alpha} = W - \tfrac{1}{2}\rho SV_0^2\frac{dC_D}{d\alpha} \\ X_\phi &= \frac{\partial X}{\partial \phi} = -W \end{aligned} \tag{8–41}$$

In a similar fashion, the force in the z-direction becomes

$$Z = -\tfrac{1}{2}\rho SV^2C_L + W \tag{8–42}$$

so that

$$\begin{aligned} Z_v &= \frac{\partial Z}{\partial v} = -\rho SV_0C_L \\ Z_\alpha &= \frac{\partial Z}{\partial \alpha} = -\tfrac{1}{2}\rho SV_0^2\frac{dC_L}{d\alpha} \\ Z_\phi &= \frac{\partial Z}{\partial \phi} = 0 \end{aligned} \tag{8–43}$$

The moment becomes

$$M = \tfrac{1}{2}\rho ScV^2C_m \tag{8-44}$$

so that

$$\begin{aligned}
M_v &= \frac{\partial M}{\partial v} = \rho ScV_0C_m \\
M_\alpha &= \frac{\partial M}{\partial \alpha} = \tfrac{1}{2}\rho ScV_0^2\frac{\partial C_m}{\partial \alpha} \\
M_{\dot\alpha} &= \frac{\partial M}{\partial \dot\alpha} = \tfrac{1}{2}\rho ScV_0^2\frac{\partial C_m}{\partial \dot\alpha} \\
M_{\dot\phi} &= \frac{\partial M}{\partial \dot\phi} = \tfrac{1}{2}\rho ScV_0^2\frac{\partial C_m}{\partial \dot\phi}
\end{aligned} \tag{8-45}$$

The equations of motion may now be written in the form

$$\begin{aligned}
m\dot v + \rho SV_0vC_D - W\alpha + \tfrac{1}{2}\rho SV_0^2\alpha\frac{dC_D}{d\alpha} + W\phi &= 0 \\
mv\dot\theta - \rho SV_0vC_L - \tfrac{1}{2}\rho SV_0^2\alpha\frac{dC_L}{d\alpha} &= 0 \\
-mk_y^2\dot q + \rho ScV_0vC_m + \tfrac{1}{2}\rho ScV_0^2\alpha\frac{\partial C_m}{\partial \alpha} + \tfrac{1}{2}\rho ScV_0^2\dot\alpha\frac{\partial C_m}{\partial \dot\alpha} \\
+ \tfrac{1}{2}\rho ScV_0^2\dot\phi\frac{\partial C_m}{\partial \dot\phi} &= 0
\end{aligned} \tag{8-46}$$

Dividing the first two equations by ρSV_0^2 and the last by $\tfrac{1}{2}\rho SV_0^2c$ and denoting

$$\frac{m}{\rho SV_0} = \tau \tag{8-47}$$

we have

$$\begin{aligned}
\frac{\tau\dot v}{V_0} + \frac{v}{V_0}C_D - \frac{W}{\rho SV_0^2}\alpha + \tfrac{1}{2}\alpha\frac{dC_D}{d\alpha} + \frac{W}{\rho SV_0^2}\phi &= 0 \\
\frac{\tau v\dot\theta}{V_0} - \frac{v}{V_0}C_L - \tfrac{1}{2}\alpha\frac{dC_L}{d\alpha} &= 0 \\
-\frac{2\tau k_y^2}{V_0c}\dot q + 2\frac{v}{V_0}C_m + \alpha\frac{\partial C_m}{\partial \alpha} + \dot\alpha\frac{\partial C_m}{\partial \dot\alpha} + \dot\phi\frac{\partial C_m}{\partial \dot\phi} &= 0
\end{aligned} \tag{8-48}$$

A dimensionless variable s will be defined as

$$s = \frac{t}{\tau} \tag{8-49}$$

and differentiation with respect to s will be indicated by a prime. We further define

$$\frac{v}{V_0} = u, \quad \frac{\dot{v}}{V_0} = \dot{u} \tag{8-50}$$

and note that

$$C_L = \frac{W}{\frac{1}{2}\rho S V_0^2} \tag{8-51}$$

Again using the subscript notation for differentiation, the equations of motion in the x and z directions may be written in the nondimensional form

$$u' + C_D u + \tfrac{1}{2}(C_{D_\alpha} - C_L)\alpha + \tfrac{1}{2}C_L\phi = 0 \tag{8-52a}$$

$$\phi' - \alpha' - uC_L - \tfrac{1}{2}C_{L_\alpha}\alpha = 0 \tag{8-52b}$$

The moment equation can be made nondimensional in a similar manner and written as

$$-\frac{2k_y^2}{\mu c^2}\phi'' + 2C_m u + C_{m_\alpha}\alpha + C_{m_{\alpha'}}\alpha' + C_{m_{\phi'}}\phi' = 0 \tag{8-52c}$$

where

$$\mu = \frac{m}{\rho S c} \tag{8-53}$$

and is called the *density factor*.

8-3. Longitudinal Stability Derivatives. The various stability derivatives appearing in Eq. 8-52, which are the final form of the longitudinal equations of motion, will now be discussed briefly (for additional details, see References 1-4).

The three derivatives C_{L_α}, C_{D_α}, C_{m_α} represent the slopes of the lift, drag, and moment curves, with respect to angle of attack. The latter of these is particularly important in that it is the criterion for *static* stability, a negative slope of the moment curve versus angle of attack being the stable condition.

The rate of change of moment with ϕ' is indicative of the damping in pitch of the airplane. While all components of the airplane add to the damping, the contribution of the horizontal tail is the most important. The value of this derivative, considering only the tail, is usually given as

$$C_{m_{\phi'}} = -\frac{a_t}{\mu}\left(\frac{S_t}{S}\right)\left(\frac{l}{c}\right)^2 \eta_t \tag{8-54}$$

where a_t is the tail-lift curve slope, S_t is the horizontal tail area, and η_t (tail efficiency factor) is the ratio of the dynamic pressure at the tail to the dynamic pressure in the free stream.

The rate of change of moment with α' accounts for the time lag required for the wing downwash to reach the horizontal tail. This derivative is usually given as

$$C_{m_{\alpha'}} = -\frac{a_t}{\mu}\left(\frac{S_t}{S}\right)\left(\frac{l}{c}\right)^2 \frac{d\epsilon}{d\alpha} \tag{8–55}$$

where ϵ is the downwash angle.

More detailed discussions and analyses of the longitudinal stability derivatives are available (References 1–4) and should be studied by the interested student. It may be mentioned, however, that the precise calculation of these derivatives is an extremely difficult task and may often not prove reliable for certain types of aircraft. Consequently, techniques for obtaining the stability derivatives by means of actual flight tests have been developed (References 4–6) and found to be very valuable, not only for obtaining their values for particular aircraft, but also for trend studies. The derivatives are also frequently evaluated from wind tunnel tests (Reference 3).

8–4. Solutions of the Equations of Longitudinal Motion. The problem now remains to solve the equations of motion (Eq. 8–52). As usual, we assume solutions of the form

$$u = Ue^{\lambda s}, \quad \alpha = Ae^{\lambda s}, \quad \phi = \Phi e^{\lambda s} \tag{8–56}$$

These trial solutions lead to three homogeneous algebraic equations, the determinant of the coefficients of which must vanish if the trial solutions are to exist. Thus

$$\begin{vmatrix} C_D + \lambda & \frac{1}{2}(C_{D_\alpha} - C_L) & \frac{1}{2}C_L \\ -C_L & -(\lambda + \frac{1}{2}C_{L_\alpha}) & \lambda \\ 2C_m & (C_{m_\alpha} + \lambda C_{m_{\alpha'}}) & \lambda\left(C_{m_{\phi'}} - \dfrac{2k_y^2}{\mu c^2}\lambda\right) \end{vmatrix} = 0 \tag{8–57}$$

The expansion of this determinant leads to a quartic equation in λ of the form

$$\lambda^4 + a_3\lambda^3 + a_2\lambda^2 + a_1\lambda + a_0 = 0 \tag{8–58}$$

as was obtained earlier by the simplified analysis (Eq. 8–21). From this equation the stability characteristics of the airplane may be determined immediately, again as was done in the simplified analysis presented earlier. The quartic equation will lead to four values of λ. If all of the λ's are real and positive, the motion will be a divergent aperiodic (non-oscillatory) one, while if the values are all real and negative, the motion will be a convergent aperiodic one. If any two of the λ's appear as complex conjugates, the motion will be oscillatory. The oscillatory motion will be damped if the real parts are negative, and divergent if the real parts are positive. Thus, if the stability conditions (Eq. 8–23) are satisfied, we see that there are two different ways, physically, in which the motion can be stable. The stable aperiodic motion corresponds to the boundary line in the stability chart shown in Fig. 8–2.

If the values of the arbitrary constants in the general solution

$$u = U_1 e^{\lambda_1 s} + U_2 e^{\lambda_2 s} + U_3 e^{\lambda_3 s} + U_4 e^{\lambda_4 s} \tag{8–59}$$

are known at the beginning of the motion ($s = 0$), it is possible to study the time history of the particular motion.

A study of the modes of motion given by the stability determinant (Eq. 8–57) will again show the existence of phugoid and short-period oscillations, but we need not undertake this task here.

A great deal of very valuable information relative to power-off longitudinal stability is given in Reference 7, including a large number of charts useful for design purposes.

8–5. Equations of Lateral Motion. Proceeding now to the case of antisymmetrical, or lateral, motions of an airplane, it is necessary that the notation be established. In addition to the angular velocity in pitch q, we denote the angular velocity in yaw by r and the angular velocity in roll by p. Similarly, in addition to the external pitching moment M, we have the external yawing moment N and the external rolling moment L. The moment of inertia about the x-axis will be denoted by I_x and that about the z-axis by I_z, while the product of inertia will be denoted by I_{xz}.

The yawing and rolling angles β and γ are defined as shown in Fig. 8–5. The angle between the x-axis at any time t and its direction at $t = 0$ is defined as ψ. The sideslip velocity, i.e., the velocity in the y direction, will be denoted by v.

By application of Newton's laws of motion, the equations of lateral motion in power-off flight may be written as

$$\begin{aligned} Y &= m(\dot{v} + Vr) \\ L &= \dot{p}I_x - \dot{r}I_{xz} \\ N &= \dot{r}I_z - \dot{p}I_{xz} \end{aligned} \tag{8–60}$$

As we did in Art. 8–2 for the longitudinal equations, the forces and moments may be expanded in terms of stability derivatives and the final equations reduced to nondimensional form. Because the process is lengthy, but follows the principles already outlined in the study of the longitudinal equations, we shall give here only the final form of the

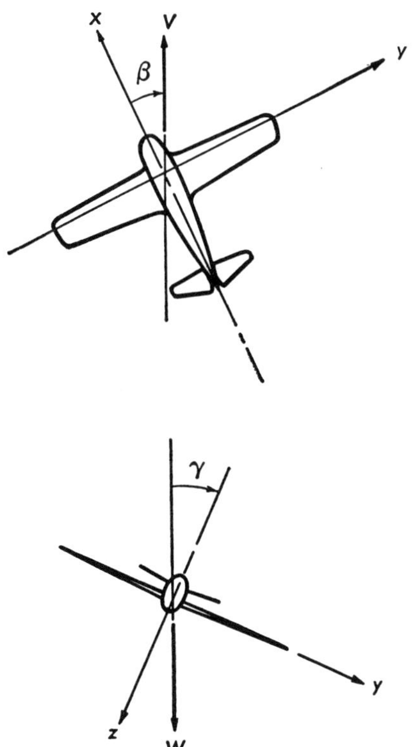

FIG. 8-5. Yawing and rolling displacements.

equations (for details see References 1–3):

$$\begin{aligned} C_{y_\beta}\beta - 2\beta' - 2\psi + C_L\gamma &= 0 \\ \mu C_{l_\beta}\beta + \tfrac{1}{2}C_{l_r}\psi + \tfrac{1}{2}C_{l_p}\gamma' - 2\left(\frac{k_x}{b}\right)^2\gamma'' &= 0 \\ \mu C_{n_\beta}\beta + \tfrac{1}{2}C_{n_r}\psi + \tfrac{1}{2}C_{n_p}\gamma' - 2\left(\frac{k_z}{b}\right)^2\psi' &= 0 \end{aligned} \tag{8-61}$$

In these equations μ has the same definition as before, primes indicate differentiation with respect to s as before, and b represents the total wing span. k_x and k_z are the radii of gyration about the x- and z-axes, respectively, and the effects of the product of inertia have been omitted entirely.*

* This is a good approximation for many airplanes because the z-axis is very nearly a principal axis. Certain high-density aircraft, however, have stability characteristics depending strongly on the products of inertia. Complete discussions are given in References 4, 8, and 9.

The stability derivatives depending upon β, i.e., C_{y_β}, C_{l_β}, and C_{n_β}, are known collectively as the static stability derivatives, and individually as the side force derivative, the directional stability derivative, and the dihedral effect derivative, respectively. These derivatives may be obtained from wind tunnel or flight tests, or they may be estimated by means of various analyses (see References 6 and 9).

The derivative C_{l_p} represents the damping in roll and may be obtained from the relation

$$C_{l_p} = \frac{2a_w}{SbV_0} \int_0^{b/2} cy^2 \, dy \tag{8-62}$$

Charts for this derivative as a function of aspect ratio and wing planform geometry are given in Reference 10. The derivative C_{n_r} represents the damping in yaw and is similar to the longitudinal derivative C_{m_q}. It can be estimated from the equation

$$C_{n_r} = -2a_v \frac{S_v}{S} \left(\frac{l_v}{b}\right)^2 \eta_t \tag{8-63}$$

where the subscript v refers to the vertical tail.

The derivatives C_{l_r} and C_{n_p} represent the coupling action involved in lateral motion and are often referred to as the "cross" derivatives. The former represents the rolling moment due to yawing velocity and is given approximately by

$$C_{l_r} = \frac{C_L}{4} \tag{8-64}$$

while the latter represents the yawing moment due to rolling velocity and is given approximately by

$$C_{n_p} = -\frac{C_L}{8} \tag{8-65}$$

8-6. Solutions of the Equations of Lateral Motion. By exactly the same procedures outlined in the case of longitudinal motion, the stability determinant may be formed as

$$\begin{vmatrix} C_{y_\beta} - 2\lambda & -2 & C_L \\ \mu C_{l_\beta} & \frac{1}{2}C_{l_r} & \frac{1}{2}C_{l_p}\lambda - 2\left(\frac{k_x}{b}\right)^2 \lambda^2 \\ \mu C_{n_\beta} & \frac{1}{2}C_{n_r} - 2\left(\frac{k_z}{b}\right)^2 \lambda & \frac{1}{2}C_{n_p}\lambda \end{vmatrix} = 0 \tag{8-66}$$

which again leads to a quartic equation of the form

$$\lambda^4 + a_3\lambda^3 + a_2\lambda^2 + a_1\lambda + a_0 = 0 \tag{8-67}$$

For a particular airplane, the stability derivatives may be evaluated and the coefficients in Eq. 8–67 determined. The values of the coefficients then lead directly to a study of the stability characteristics of the airplane. As we have seen previously, stability will be ensured if all the coefficients are positive and if the inequality

$$a_1^2 - a_3(a_1a_2 - a_0a_3) < 0 \tag{8–68}$$

is satisfied.

Studies of the general quartic equation (Eq. 8–67) show that if the coefficient a_0 is negative, there will be one positive real root of the equation and hence a divergent motion. In the present case

$$a_0 = C_{l\beta}C_{n_r} - C_{n_\beta}C_{l_r} \tag{8–69}$$

and for most aircraft this turns out to be a negative value, with the resulting real positive root of the stability quartic. The remaining roots are almost always a real negative root and a pair of complex conjugate roots.

The divergent mode of motion is called *spiral divergence* and the corresponding spiral boundary is found by putting Eq. 8–69 equal to zero. The other nonoscillatory mode is of no particular consequence, since it is always highly convergent. The lateral short-period oscillatory motion can be of some importance, particularly if it is lightly damped—in such a case it is referred to as *Dutch roll* motion. The oscillatory stability boundary can be defined by making Eq. 8–68 an equality with zero, rather

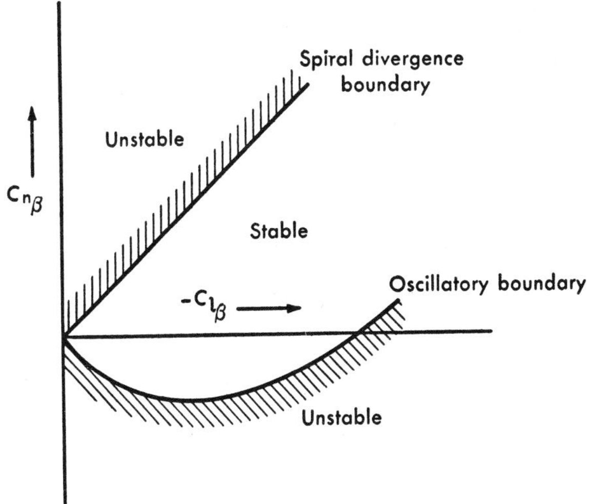

FIG. 8–6. Lateral stability chart.

than an inequality. Typical stability boundaries are shown in Fig. 8–6. The usual case is for an airplane to have the short-period oscillation fairly highly damped but to possess a slight spiral instability.

A very valuable summary of lateral-stability information, with useful design charts, is given in Reference 11.

8–7. Additional Topics. FREE CONTROLS. The previous discussions of the equations of motion of a rigid airplane have not considered the influences of freely moving control surfaces; that is, all control surfaces have been assumed to be locked in position.* Actually, of course, these surfaces are always free to move to a greater or lesser degree—in some instances this may be the result of the pilot flying "hands-off" or it may be due to the slight amount of slack motion present in almost every control system.

For longitudinal motion, the appropriate control surface is the elevator. Consideration of the elevator degree of freedom results in a fourth equation of motion as well as coupling terms in the others, with consequent definition of several new stability derivatives dependent upon the characteristics of the particular type of control surface. The resulting stability determinant is of fourth order, and it is found that there are three possible modes of motion—the phugoid and two short-period motions. The phugoid is usually found to be influenced only very slightly by a free elevator, and hence it is possible to eliminate the equation of motion in the x direction and thus reduce the problem to one of third order in the determinant, as before. These investigations are beyond the scope and purpose of this book, but we may profitably discuss briefly the modes of motion that are obtained.

The two modes of stick-free motion occur at essentially constant airspeed and have short periods. The first is essentially the free rotation of the elevator about its hinge line and is usually of very high frequency. The other mode is of somewhat longer period (say, 2 to 3 sec) but is still far less than the period of the phugoid, which may be on the order of a minute or two. If this mode is only slightly damped the motion is known as *porpoising*.

Of the four modes of longitudinal motion that have been mentioned, the short-period stick-fixed and stick-free modes are usually quite heavily damped and therefore are not of major concern. The phugoid motion, even when lightly damped or even slightly unstable, is within the ability of the pilot to control because of its long period and therefore can be tolerated to some extent. The short-period stick-free mode (porpoising) is the most undesirable and often can be quite dangerous.

* This type of analysis is often referred to as the "stick-fixed" analysis. The analysis considering free control surfaces is referred to as "stick-free."

In the case of antisymmetric (lateral) motion, the situation is more complicated because there are two control surfaces to be considered, rudder and ailerons. The lateral stability problem with free controls then becomes one of five degrees of freedom with the associated increase in the number of stability derivatives to be evaluated. The analysis then becomes quite complex because the coupling effects are so involved. For simplicity in discussion, then, it is often convenient to consider the case wherein only the rudder is free and then the case where only the ailerons are free. Even so, we must deal with a fourth-order system, the stability equation being of sixth degree in λ.

In the free-rudder case, the six roots lead to four modes of motion. Two of these are aperiodic motions, one of which corresponds to the stick-fixed convergent rolling mode and the second of which is a slightly divergent spiral mode of little importance. The other two modes are oscillatory. One of these is a simple, short-period, highly damped oscillation of the rudder about its hinge line. The second oscillatory mode is also of short-period and is predominately a yawing motion which has been termed *snaking*.

The free-ailerons case again leads to a sixth-degree stability equation, but in this instance five modes of motion are usually found. One of these is the slightly divergent spiral mode, three others are convergent aperiodic modes of no particular importance, and the final mode is the stick-fixed, Dutch roll oscillation obtained previously.

The effect of friction in the control systems has been studied, and it has been found that, under certain circumstances, steady oscillations which are highly undesirable may result (Reference 12).

Power-on Flight. It is probably worthwhile to consider briefly some of the effects of power on the stability analyses. In the case of jet aircraft, it is necessary only to add a thrust term to the equations of motion, since there is practically no influence on the various factors which govern stability. In the case of propeller driven aircraft, however, the effects are quite pronounced and are extremely difficult to evaluate. These may be broken down into three groups: (1) slipstream effects, (2) propeller fin effect, and (3) effects on stability derivatives.

The propeller imparts an additional velocity to a certain portion of the air passing over the airplane, and hence the portion of the tail (horizontal or vertical) immersed in the slipstream is increased in effectiveness. The slipstream also increases the downwash angle at the tail, and thus offers a destabilizing effect.

A rotating propeller acts somewhat like a stationary airfoil when the plane of rotation is not normal to the relative wind and the lift vector lies in the plane defined by the relative wind and the normal. Thus, when the airplane pitches, the propeller acts partly in the same fashion as a lifting surface. The result, because of the forward location of the force and the resulting downwash, is to cause some reduction in stability.

Finally, the propeller influences certain of the stability derivatives to a marked degree, and wind tunnel tests are usually conducted in order to evaluate these effects.

High-Speed Flight. It is well known that as flight speeds increase, the Mach number enters aerodynamic analyses as a new variable. The equations of longitudinal and lateral motion developed previously in this chapter, as well as the methods of solution and methods of investigating stability, are valid for all flight speeds. The only restriction is that of linearization. Therefore, if these results are to be applied to flight at high subsonic or supersonic speeds, it is necessary only to obtain values of the stability derivatives and other coefficients valid at these speeds. When the influence of compressibility is to be considered, the various force and moment expansions in terms of derivatives are no longer independent of the flight speed, and consequently, new terms appear in the expressions for certain of the stability derivatives. The derivatives are also modified by changes in the coefficients and in the rates of change of the derivatives with angle of attack at constant Mach number. Detailed information relative to some of the resulting changes in stability characteristics are given in References 3, 13, and 14.

Response Calculations. Thus far we have discussed only solutions of the equations of motion in the absence of external exciting forces and moments. Such external forces and moments might arise, for example, from the effect of a gust having a certain, specified time history, or from certain prescribed movements of the controls. The complete solution of the equations of motion then consists of the general solutions obtained for the homogeneous equations plus the particular solutions corresponding to the external forcing functions. Exact solutions obtained by this method are extremely complicated and tedious even when the forcing functions are extremely simple (Reference 1).

If the forcing function is arbitrary and cannot be expressed by a simple mathematical function, the only recourse is to numerical methods of solution, which of course demand almost prohibitive expenditures of labor. Another slightly simpler procedure is to obtain the solution for a linear forcing function and then to represent the arbitrary forcing functions as a series of linear increments and add the solutions in a suitable manner. The method of the Laplace transform is often quite useful for simple forcing functions (Reference 15) and has the advantage of automatically including the initial conditions so that the arbitrary constants for the free vibrations are directly evaluated. Finally, it is possible to represent the arbitrary forcing function by a Fourier series, but this method too is often prohibitive if the number of terms required for an adequate representation is not small.

As a very simple example of a response calculation, let us consider longitudinal motion at constant airspeed when the elevator is suddenly displaced. Eq. 8–52 then reduce to

$$\begin{aligned} \alpha' + \tfrac{1}{2}C_{L_\alpha}\alpha - \phi' &= 0 \\ \frac{-2k_y^2}{\mu c^2}\phi'' + C_{m_{\phi'}}\phi' + C_{m_{\alpha'}}\alpha' + C_{m_\alpha}\alpha &= M_\delta\delta_e(t) \end{aligned} \qquad (8\text{–}70)$$

where M_δ represents the amplitude of the pitching moment produced by the elevator defection δ_e. These two equations may be combined by differentiating the first so that the remaining unknown is α:

$$\begin{aligned} \alpha'' + [\tfrac{1}{2}C_{L_\alpha} - K(C_{m_{\alpha'}} + C_{m_{\phi'}})]\alpha' - K(C_{m_\alpha} + \tfrac{1}{2}C_{L_\alpha}C_{m_{\phi'}})\alpha & \\ = KM_\delta\delta_e(t) & \end{aligned} \qquad (8\text{–}71)$$

where

$$K = \frac{\mu c^2}{2k_y^2} \qquad (8\text{–}72)$$

This equation is equivalent to that for a simple single-degree-of-freedom system, and hence can be solved quite readily for α, where we recall that this is the change in angle of attack in a slightly disturbed motion.

Another very simple example is provided by the pure rolling response to an abrupt aileron movement. The equations of Eq. 8–61 reduce to the single equation

$$\tfrac{1}{2}C_{l_p}\gamma' - 2\left(\frac{k_x}{b}\right)^2\gamma'' = -R_\delta\delta_a(t) \qquad (8\text{–}73)$$

which can also be solved quite readily by classical methods. The response to an aileron displacement when systems of two-and-three-degrees-of-freedom are considered is given in Reference 2.

8–8. Effects of Unsteady Flow and Structural Flexibilities. The analyses of this chapter thus far have been concerned entirely with rigid aircraft. In practice, structural components are often so sufficiently flexible that important types of problems such as divergence, control surface effectiveness, and flutter must be investigated, as we have seen in previous chapters. It is to be expected that structural flexibility may have appreciable influences on airplane flight stability in certain cases, and it is therefore the purpose of this article to outline some of these influences and the methods that can be employed to estimate them.

Local distortions, such as those occurring at rather discrete locations on the airplane, and particularly those of control surfaces, may be of considerable importance with reference to control forces. These may be accounted for fairly simply by employing proper values of the coefficients, as obtained from analyses, flight test data, or wind tunnel data. Deformations of major structural components, such as bending or twisting of a wing or fuselage, modify the airplane configuration, and hence modify the equations of motion to a significant extent. Further, if the deformations are dynamic, i.e., if they are functions of time, the aerodynamic analysis requires consideration of the nonsteady flow, as we have seen in Chapter 5.

One way of accounting for these effects is to write, in addition to the rigid-body equations already considered, the equations describing certain of the more important distortion modes. For example, it may be that, in addition to the rigid-body modes of longitudinal motion, fuselage bending and wing twisting are considered to be of importance. The three rigid-body equations, the elevator control equation, and the two distortion equations thus lead to a six-degree-of-freedom system. The entire system is complicated, of course, by the fact that terms must be introduced to account for the couplings between the distortion coordinates and the dynamic variables of the rigid-body airplane, in addition to unsteady-flow terms. But even this complex representation can be far too simple in many cases, i.e., in those cases where several distortion modes are required to describe the actual motions adequately.

Another method of accounting for these effects is to consider modified forms of the various stability derivatives. These modifications are usually based on static deformations; hence they are valid only when the distortions have a much lower frequency than the lowest natural frequency of the structure and when the forces and moments depending upon the distortional velocities and accelerations are negligible.

In any case, certain unsteady flow effects are of importance and must be considered. No unsteady or accelerated flow terms are usually included in the equations, and therefore we shall discuss some of the aspects of this problem briefly in the remainder of this article.

The term M_α' retained in the simplified longitudinal equations (Eq. 8–39) is a "quasi-steady" acceleration correction to allow for the time required for the downwash (vortex sheet) produced by the wing to reach the tail. All other quasi-steady and unsteady terms were omitted in the simplified equations we considered.* In order to consider some of these, we shall refer to the flutter problem and employ concepts developed in Reference 16. We consider two-dimensional flow for simplicity.

The first task is to transform the flutter axes to the stability (wind) axes used previously in this chapter (see Fig. 8–7). The flutter axes are

* The quasi-steady term Z_α is often included even in the simplified equations, depending, as it does, on the time for the downwash to reach the tail in the same way as in the case of M_α'.

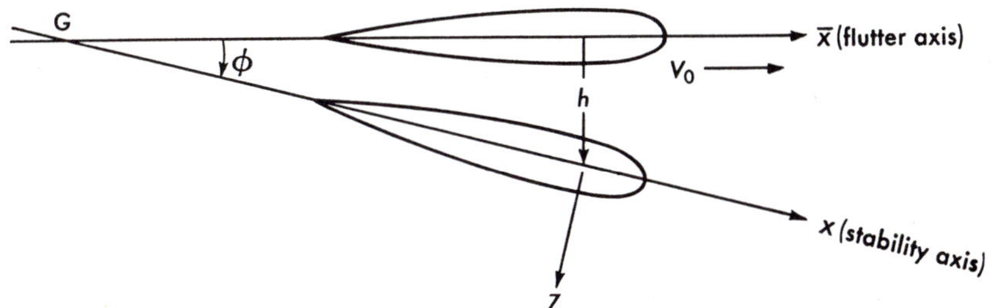

FIG. 8–7. Flutter and stability axes.

such that the $\bar{x}$-axis moves vertically with a constant linear velocity, while the wing rotates through some angle θ. The point of rotation is denoted by $\bar{x} = G$ and moves vertically a distance h from the $\bar{x}$-axis. These axes may be transformed to the wind axes (x) which are fixed in the wing and move with it, with the center of rotation at $x = G$, according to the relation

$$\alpha = \frac{w}{V_0} = \phi + \frac{\dot{h}}{V_0} \tag{8–74}$$

where w is the vertical velocity. Thus, any stability derivatives which depend on α or $\dot{\alpha}$ are determined by setting $q = \dot{\phi} = 0$, $\theta = 0$, and $w = \dot{h}$, while derivatives with respect to $\dot{\phi}$ or $\ddot{\phi}$ are determined by setting $\dot{h} = -V_0\phi$.

The resultant lift force for two-dimensional flow about an oscillating airfoil is,* referred to the flutter axis $\bar{x}$,

$$Z = -\frac{\pi}{4}\rho c^2 V_0\left[\dot{\phi} + \frac{\ddot{h}}{V_0} + \left(\frac{1}{2} - \frac{G}{c}\right)\frac{c\ddot{\phi}}{V_0}\right] - \pi\rho c V_0^2\left[\phi + \frac{\dot{h}}{V_0} + \left(\frac{3}{4} - \frac{G}{c}\right)\frac{c\dot{\phi}}{V_0}\right]C(k) \tag{8–75}$$

Upon transferring to the stability axis x, this reduces to the form

$$Z = -\frac{\pi}{4}\rho c^2 \dot{w} - \pi\rho c V_0^2\left(\frac{w}{V_0}\right)C(k) \tag{8–76a}$$

for taking derivatives with respect to $\dot{\alpha}$. For taking derivatives with respect to $\dot{\phi}$, the transformed equation takes the form

$$Z = -\frac{\pi}{4}\rho c^2\left(\frac{1}{2} - \frac{G}{c}\right)c\ddot{\phi} - \pi\rho c^2 V_0\left(\frac{3}{4} - \frac{G}{c}\right)\dot{\phi}C(k) \tag{8–76b}$$

From Eq. 8–76 it is possible to evaluate the stability derivatives Z_α, $Z_{\dot{\alpha}}$, $Z_{\dot{\phi}}$, and $Z_{\ddot{\phi}}$. Of these, only the steady-state form of Z_α has been obtained previously (see Eq. 8–43b).

* See Art. 5–4.

As previously discussed, the most important unsteady effect is the phase lag of the action of the wing downwash on the tail. Eqs. 8–76 show that $Z_{\dot{\phi}}$, as well as Z_{α}, must contribute to this downwash lag because any net change in lift ($Z_{\dot{\phi}}$) must appear as a change in downwash even though there is no change in α. This change in lift with $\alpha = 0$ results from a change in the chordwise vortex distribution, which in turn results from the effective change in wing camber due to the curvature of the streamlines corresponding to $\dot{\phi}$. The lift or downwash change due to $\dot{\phi}$ is therefore given directly by the local vertical velocity produced at the ¾-chord point. This change in downwash is usually negligible at low frequencies of vibration, but it can be very important at higher frequencies. On the other hand, at very high frequencies the wave length of the fluctuation in downwash becomes less than the chord length of the tail, and hence the effective downwash angle becomes zero and need not be considered.

The moment derivatives can be studied in a manner similar to that above for the vertical force. It is found in such an analysis that the acceleration derivatives $M_{\ddot{\alpha}}$ and $M_{\ddot{\phi}}$ may both become significant.

A brief discussion of certain effects of flexibilities on lateral stability is given in Reference 17, while some further discussions of stability derivatives in unsteady flow are given in Reference 18.

PROBLEMS

8–1. Using Eq. 8–24, plot longitudinal stability boundaries for $\alpha_0 = 0, 5, 10$ degrees.

8–2. Repeat the simplified longitudinal analysis of Art. 8–1, but include the drag as a small constant force. Confirm the fact that in a $\sigma - \delta$ diagram, the stability boundary does not pass through the origin.

8–3. Derive the nondimensional form of the moment equation for longitudinal motion as given in Eq. 8–52c.

8–4. Expand the longitudinal frequency equation (Eq. 8–57), and write the coefficients of the quartic equation.

8–5. The longitudinal stability derivatives and other parameters of a certain airplane are such that the coefficients of the stability quartic are (Problem 5, Chapter 4) $a_0 = ½$, $a_1 = 1$, $a_2 = 5.02$, $a_3 = 3.85$. Discuss the stability characteristics of this airplane, including calculations for the four roots of the quartic and the periods of the modes of motion.

8–6. Derive Eq. 8–60.

8–7. Find the complete solutions of Eqs. 8–71 and 8–73 if the control surface displacements are taken as applied suddenly and then held constant (step function).

8–8. Using Eq. 8–76a, evaluate the stability derivative $Z_{\dot{\alpha}}$, and compare with Eq. 8–43b for the case of vibration frequency equal to zero.

REFERENCES

1. JONES, B. M. "Dynamics of the Airplane," *Aerodynamic Theory*. Vol. 5. Edited by W. F. DURAND. New York: Springer Publishing Co., Inc., 1935.

2. Perkins, C. D., and Hage, R. E. *Airplane Performance Stability and Control.* New York: John Wiley & Sons, Inc., 1949.
3. Duncan, W. J. *Control and Stability of Aircraft.* New York: Cambridge University Press, 1952.
4. Milliken, W. F. "Progress in Dynamic Stability and Control Research," *Jour. Aero. Sciences,* **14** (September, 1947): 493–519.
5. Donegan, J. J., and Pearson, H. A. "Matrix Method of Determining the Longitudinal Stability Coefficients and Frequency of an Aircraft from Transient Flight Data," *NACA Tech. Note* 2370 (1951).
6. Greenberg, H. "A Survey of Methods for Determining Stability Parameters of an Airplane from Dynamic Flight Measurements," *NACA Tech. Note* 2340 (1951).
7. Zimmerman, C. H. "An Analysis of Longitudinal Stability in Power-off Flight with Charts for Use in Design," *NACA Rept.* 521 (1935).
8. Sternfield, L., and McKinney, M. O. "Dynamic Lateral Stability as Influenced by Mass Distribution," *Jour. Aero. Sciences,* **15** (July, 1948): 411–417.
9. Campbell, J. P., and McKinney, M. O. "Summary of Methods for Calculating Dynamic Lateral Stability and Response and for Estimating Lateral Stability Derivatives," *NACA Tech. Note* 2409 (1951).
10. Pearson, L. D., and Jones, R. T. "Theoretical Stability and Control Characteristics of Wings with Various Amounts of Taper and Twist," *NACA Rept.* 635 (1938).
11. Zimmerman, C. H. "An Analysis of Lateral Stability in Power-off Flight with Charts for Use in Design," *NACA Rept.* 589 (1937).
12. Greenberg, H., and Sternfield, L. "A Theoretical Investigation of Longitudinal Stability of Airplanes with Free Controls Including Effect of Friction in Control Systems," *NACA Rept.* 791 (1944).
13. Donovan, A. F., Flax, A. H., and Cheilek, H. A. "Stability and Control of Supersonic Aircraft," *Preprint No.* 136. New York: Institute of the Aeronautical Sciences, 1948.
14. Neumark, S. "Longitudinal Stability, Speed, and Height," *Aircraft Engineering* (November, 1950), 323–334.
15. Mokrzycki, G. A. "Application of the Laplace Transformation to the Solution of the Lateral and Longitudinal Stability Equations," *NACA Tech. Note* 2002 (1950).
16. Laitone, E. V., and Walters, E. R. "The Application of Nonstationary Airfoil Theory to Dynamic Stability Calculations," *Jour. Aero. Sciences,* **18** (March, 1951): 214–216.
17. Rodden, W. P. "An Aeroelastic Parameter for Estimation of the Effects of Flexibility on the Lateral Control and Stability of Aircraft," *Jour. Aero. Sciences,* **23** (July, 1956): 660–662.
18. Wood, R. M., and Murphy, C. H. "Aerodynamic Derivatives for Both Steady and Unsteady Motion of Slender Bodies," *Jour. Aero. Sciences,* **22** (December, 1955): 870–871.

CHAPTER 9

MISCELLANEOUS TOPICS

9–1. Advanced Topics and Miscellaneous Dynamics Problems. The discussions of flutter, aeroelasticity, impulsive loading, and flight stability given in the earlier chapters of this book were intended to serve only as introductions to these various aspects of the dynamics of airplanes. Lest the student feel that a more comprehensive volume on this general subject would only present more extended and detailed theoretical developments of these same problems, the present article has as its purpose the presentation of a brief discussion and description of certain advanced problems and topics in flutter analysis, aeroelasticity, and impulsive loading as well as certain miscellaneous dynamics problems.

FLUTTER. Supersonic flow past thin panels, such as wing, tail, or fuselage coverings, can under certain conditions give rise to vibrations of the panel normal to its own plane. This phenomenon is referred to as *panel flutter*. The possibility of unstable oscillations resides in the fact that in supersonic flow the pressure is directly proportional to the slope of the surface (linearized theory). Thus, a symmetrical slope configuration of the surface leads to an unsymmetrical aerodynamic loading, tending to deform the surface to a more complicated shape. As the dynamic pressure is increased, it is found that a buckled plate with fixed edges cannot maintain any equilibrium configuration, and hence there exists an aerodynamic instability. The problems of wind-generated surface waves on lakes and oceans and the fluttering of flags are closely related to panel flutter.

As a very simple example (Ref. 1) of a formulation of the panel flutter problem, consider an infinitely wide panel of length $2b$, rigidly fixed along the leading and trailing edges, past one side of which is flowing a two-dimensional supersonic stream (Fig. 9–1). For simplicity, we treat the panel as an elastic membrane having a mass m per unit area and a tension T per unit width in the x direction. The excess of pressure caused by the deflection of the membrane will be denoted by $P(x)$ and the motion of the panel during flutter will be given by

$$z(x, t) = Z(x)be^{i\omega t} \tag{9–1}$$

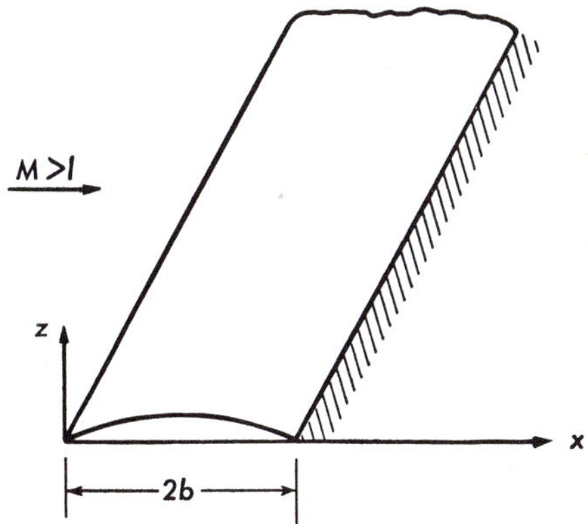

FIG. 9–1. Two-dimensional panel in supersonic flow.

where $Z(x)$ is a function of x alone and ω is the flutter frequency. The equation of motion then can be written as

$$T\frac{d^2Z}{dx^2} + m\omega^2 Z = P(x) \tag{9–2}$$

If Z and x are made nondimensional with b, this equation becomes

$$\frac{d^2Z}{dx^2} + \frac{\pi^2}{4}\left(\frac{k}{k_n}\right)^2 Z = \frac{b}{T}P(x) \tag{9–3}$$

where

$$k = \frac{\omega b}{V} \tag{9–4}$$

We could also formulate this problem in terms of a thin, simply supported plate. The corresponding equation of motion would then be

$$\frac{d^4Z}{dx^4} - \frac{\pi^4}{16}\left(\frac{k}{k_n}\right)^2 Z = -\frac{b^3}{D}P(x) \tag{9–3a}$$

where D is the plate flexural rigidity.

The pressure $P(x)$ can be obtained from an aerodynamic analysis and written in the form

$$P(x) = \frac{\rho V}{\sqrt{M^2 - 1}}\left[w(0)G(x) + \int_0^x G(x - \xi)\left(\frac{dw}{d\xi} + ikw\right)d\xi\right] \tag{9–5}$$

where $w(x)$ is the downwash, given by

$$w = V\left(\frac{dZ}{dx} + ikZ\right) \tag{9–6}$$

and

$$G(x) = J_0(\kappa x)e^{-i\kappa Mx} \tag{9–7}$$

where J_0 is the Bessel function of the first kind and zero order, while

$$\kappa = \frac{mk}{M^2 - 1} \tag{9-8}$$

The final equation of motion, considering the membrane formulation, can be written as

$$\begin{aligned}\frac{d^2Z}{dx^2} + \alpha^2 k^2 Z = \gamma \Bigg\{ \left(\frac{dZ}{dx}\right)_{x=0} G(x) \\ + \int_0^k G(x - \xi)\left[\frac{d^2Z}{d\xi^2} + 2ik\frac{dZ}{d\xi} - k^2 Z\right] d\xi \Bigg\}\end{aligned} \tag{9-9}$$

where

$$\begin{aligned}\alpha &= \frac{\pi}{2} k_n \\ \gamma &= \frac{\pi^2}{4\sqrt{M^2 - 1}} \frac{\rho b}{m k_n^2}\end{aligned} \tag{9-10}$$

The solution of this formulation of the panel flutter problem then consists of solutions of Eq. (9–9) that also satisfy the boundary conditions $Z = 0$ at $x = 0, 2$.

Other analyses of the panel flutter problem, which consider panels on multiple supports, various edge conditions, change in length of the panel due to axial thrust, and thermal buckling effects, may be found in Refs. 1–3.

Let us now consider briefly certain other advanced aspects of flutter.

For certain ranges of Mach number and for certain locations of the axis of rotation, it has been found that flutter in a single-degree-of-freedom mode is possible. Of the various types of single-degree-of-freedom flutter studied, such as pitching motion of an unswept wing, bending motion of a swept wing, or control surface oscillation, the pitching motion of an unswept wing is one of the most interesting. The equation of motion is obtained readily from the results of Chapter 5 as

$$I_\alpha \ddot{\alpha} + (1 + ig_\alpha) C_\alpha \alpha = M_\alpha \tag{9-11}$$

Studies of this equation have shown (Ref. 4) that unstable pitching oscillations are possible, even at very low speeds, if the axis of rotation is located well forward of the wing. A comparison between theory and experiment is shown in Fig. 9–2.

Frequently, as the angle of attack of a wing is increased to values near the stalling angle, it is observed that the flutter velocity is much lower than when the wing is near the zero-lift angle. The classical flutter theory based on potential flow cannot account for the effects of separated flow, and therefore flutter in this range of lift coefficients, even though the wing may not be completely stalled in the usual sense, is called *stall flutter* (Refs. 5 and 6). This type of flutter is usually predominantly a single-degree-of-freedom motion in the torsion mode. Since airplanes and missiles rarely attain high-lift coefficients in high-speed flight, the stall flutter problem is not particularly important from the standpoint of wing or tail surfaces; on the other hand, high-speed stalling of rotors and propellers is relatively common, and under certain conditions it is quite possible for compressor and turbine blades to undergo a type of stall flutter instability.

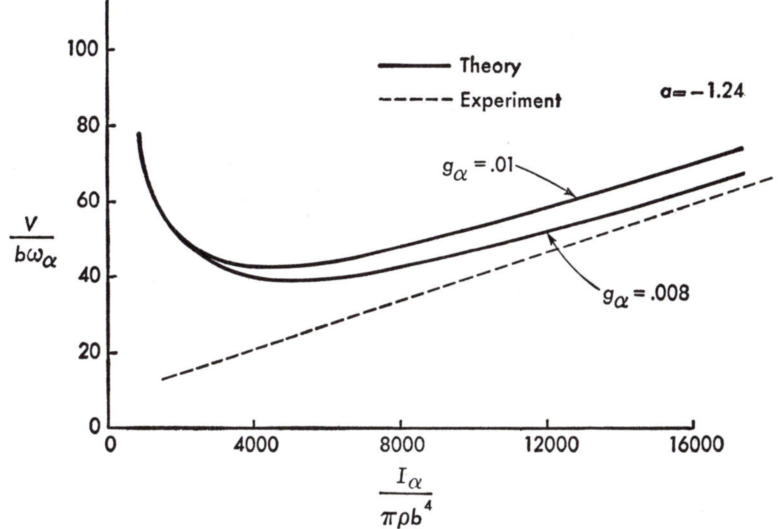

FIG. 9–2. Single-degree-of-freedom flutter.

In addition to the two principal characteristics of stall flutter already mentioned, i.e., a decrease in flutter speed and a predominantly torsional mode of oscillation, the following characteristics are of considerable interest. As the flutter speed decreases, a minimum value is reached and rises again as the surface becomes completely stalled. In this latter range the motion more nearly corresponds to forced vibration than to self-excited flutter; the reason for this is that the wake frequency (Kármán vortex street) becomes well defined, and the surface reacts to the changes in circulation which result from the vortex formation according to the Kutta-Joukowski law. This situation is not, however, the same as buffeting, since there the vibration is forced by the vortexes impinging on a surface located in the wake. In the completely stalled range of stall flutter, the amplitude is usually quite small compared with the large-amplitude violent flutter occurring in the range of partial stall. Also, the phase relationship between bending and torsion is substantially different in stall flutter from that in classical flutter, and it is found that variations in structural properties have very different effects in stall flutter than in classical flutter. The phenomenon is essentially a nonlinear one, and consequently, analytical study of the problem is extremely difficult.

STATIC AEROELASTICITY. Insofar as more advanced topics of aeroelasticity are concerned, it may suffice to mention briefly the possibility of aeroelastic reversal of propeller blades (Ref. 7), and the divergence of supersonic wings (Ref. 8).

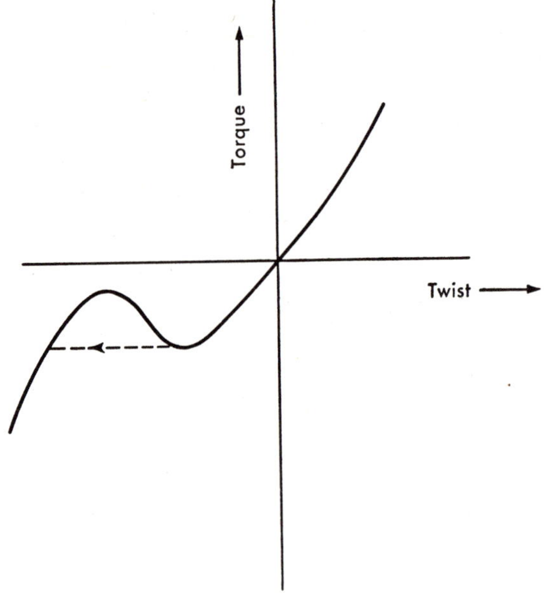

FIG. 9–3. Propeller characteristic curve.

Under certain conditions a propeller blade may have a torque-deflection characteristic of the form shown in Fig. 9–3. If a propeller blade has such a characteristic, the deflection may take a sudden jump while negative moment is being applied, as shown by the dotted line in Fig. 9–3. Similarly, on unloading, the twist may suddenly decrease at constant torque. If the blade is to be statically stable, the characteristic curve should have a positive slope everywhere.

In supersonic flight, with the consequent use of very thin, low aspect ratio wings, the static aeroelastic problem takes on an essentially more complex aspect. This is due in large measure to the fact that for such wings the assumption of high structural rigidity in the chordwise direction is not reasonable. The analysis of Ref. 8 shows that consideration of chordwise deformations leads to stability relations which depend not only on the aspect ratio, but also on the ratio of the spanwise bending stiffness at the mid-chord to the chordwise bending stiffness, and on the value of Poisson's ratio (as this quantity governs the degree of the anticlastic curvature). A further complication is introduced by the possibility of conditions whereby it is possible for the leading edge of a supersonic wing to exhibit an instability of its own—that is, a "curling-up" of the leading edge.

Impulsive Loading. A problem of impulsive loading of considerable importance involves the penetration of a moving gust.* The penetration of a blast wave, for example, would represent a similar problem because the front of the disturbance is not stationary with respect to the surrounding air but moves along the flight path—the aircraft may intercept the disturbance front or may overtake it. The unsteady lift functions for such problems must be more complicated than either Küssner or Wagner functions because they must contain the rate of travel of the gust front as an additional parameter.

For very strong blasts originating at a great distance from the aircraft, the blast-induced speed of the disturbed air behind the blast wave decreases so slowly in comparison with the time of penetration that the problem may be treated as that of penetration of a sharp-edged moving gust. For less intense blasts originating close to the aircraft, the speed of the disturbed air behind the front may decrease sufficiently rapidly to allow calculation of the lift responses to sharp-edged gusts.

A response problem of great importance, which has been discussed only very briefly in this book, is that of buffeting. As mentioned earlier in this article, the problems of stall flutter and buffeting have certain features of similarity; they may, however, occur at the same time in a manner which prevents their division into two distinctly separate phenomena. A typical plot of buffet and stall-flutter boundaries is shown in Fig. 9–4, while typical time histories for buffeting and stall flutter are shown in Fig. 9–5. The following definitions have been offered (Ref. 5) to distinguish between the two phenomena: When the point is reached where the boundary layer separation becomes unstable, there is a continuous excitation of the structure by the aerodynamic forces resulting from the separation of the air flow, and the amplitude of the response depends upon the damping forces. If the damping is very small or negative, the vibration may be termed *stall flutter*; if the damping is relatively large and always positive so that random responses are exhibited, the vibration may be termed *buffeting*.

* See, e.g., Ref. 32 of Chapter 7.

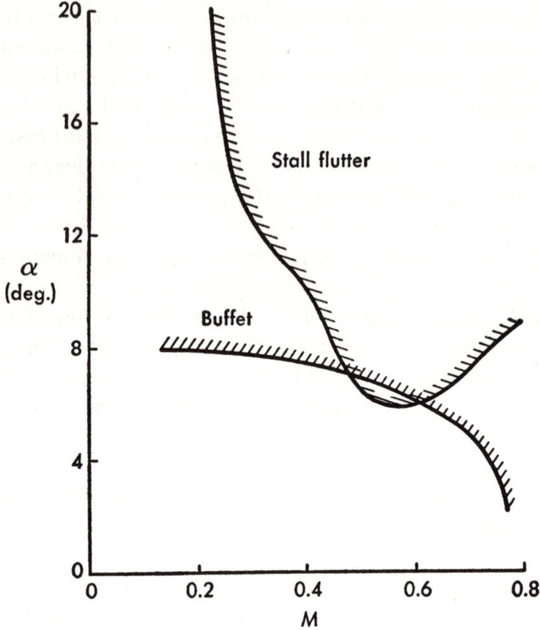

FIG. 9–4. Stall flutter and buffet boundaries.

The foregoing distinction between buffeting and stall flutter is useful for distinguishing between the two, but the *fundamental* difference between the two should always be kept in mind—buffet is a forced vibration, while stall flutter is a self-excited vibration.

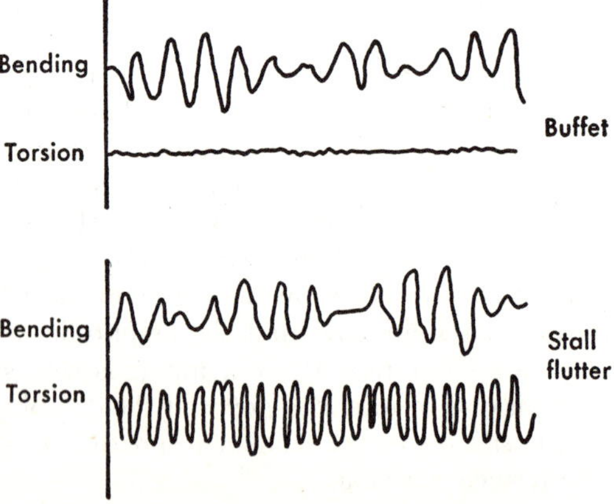

FIG. 9–5. Stall flutter and buffet time histories.

FLIGHT STABILITY. The problem of flight stability has taken on ever-increasing importance as airplanes and missiles for high-speed flight have been developed. Consequently, emphasis has been placed on more precise derivations of the equation of motion, which accounts for large disturbance motions as well as for other factors of concern to the analyst, such as reference to a set of axes fixed in the earth and steady rolling motion. For example, special studies of the equations of motion of missiles with reference to the kinematical relationships have been given in Refs. 9 and 10. A special study of the dynamic motion of a missile descending through the atmosphere and undergoing oscillatory motion about its center of gravity has been made in Ref. 11. Special studies of missiles in steady, rolling flight have been given in Refs. 10 and 12. In this latter case the problem is greatly complicated by the fact that the introduction of a rolling velocity into the analysis also introduces coupling terms which prevent the separation of the general motion into independent lateral and longitudinal motions.

New aircraft configurations have necessitated reinvestigation of the stability and control characteristics of airplanes. For example, in Ref. 13 there is reported an investigation of the response of a swept wing, tailless airplane to control movements in landing approaches as compared with that of more conventional aircraft.

MISCELLANEOUS DYNAMICS PROBLEMS. While the preceding discussions have related to certain more or less advanced topics in the various areas of airplane dynamics treated in the main part of this book, mention should also be made of a few of the multitude of specialized problems of airplane dynamics. It is to be hoped that these few selected problems will indicate, rather than completely survey, the various types of special problems with which the worker in the field of airplane dynamics may find himself confronted.

The development of high-performance airplanes and missiles has given rise to a number of important problems in stability and control. In order to overcome the resulting large stick forces, power controls and booster systems have had to be introduced, and in order to improve the precision of flight path control, automatic systems have been introduced to supplement the human pilot. The ultimate has been achieved in the guided missile, with complete replacement of the pilot by automatic systems. Thus, analytical methods must be extended to permit the calculation of the motion of an aircraft when under the control of an automatic pilot. The analysis of automatic control systems may be accomplished by the well-established techniques of servomechanism theory, but the analysis of the aircraft under automatic control and the synthesis of systems to stabilize and control the aircraft under all flight conditions can be achieved only by combining the airplane dynamic and servomechanism problems.

There is every reason to expect further that as performance characteristics of aircraft increase, the dynamic coupling between the aeroelastic modes of the aircraft and the response characteristics of the control system will become of increasing importance. An excellent summary of the analytical and experimental methods that are used in the design of aircraft automatic-control devices is given in Ref. 14, while studies involving the integration of the airplane and the control system are given in Refs. 15 and 16. A procedure for treating the complete aeroelastic, automatically controlled aircraft or missile is given in an article by Rea, Ref. 17.

The design and operation of helicopters has brought forth a number of new and interesting problems of special interest to the dynamics specialist. For example, the calculation of rotor-blade response to various loading conditions has been accomplished by solution of the appropriate equations of motion, by either tabular or matrix methods (Ref. 18). It has been observed that the stability of the transient flapping motion of the rotor blades is not constant but varies with the advance ratio. This flapping instability is quite important in helicopter operations and has been analyzed aeroelastically on the basis of two degrees of freedom (rigid flapping and elastic bending) in Ref. 19.

Undesirable rotor vibrations have occurred in helicopters with hinged blades even when not in flight but simply during power application while on the ground. The energy source in this case is provided by the rotation of the rotor rather than by the airforces (Ref. 20). The several types of vibration that may occur during flight operation can be classified according to the manner of excitation. At low rotational speeds a hinged rotor will vibrate under the stimulus of external forces as will any other dynamic system. If the dynamical system, the airforces, or any other forces or combinations of these things are such that the system exhibits negative damping, the system undergoes self-excited vibrations. On the other hand, when a rotating system is not perfectly balanced, the centrifugal force of the unbalanced mass may excite vibrations which exhibit resonance at certain rotational speeds; these are called *shaft-critical* vibrations. In this type of vibration the stiffness of the pylon is balanced by the centrifugal force at certain rotational speeds. An analysis of self-excited and shaft critical vibrations of hinged rotor blades, considering the hinge deflection of the blades in the plane of rotation and the horizontal deflections of the pylon as the appropriate degrees of freedom, is given in Ref. 20.

Stability problems of helicopter flight are also of great importance. As with conventional aircraft, the long period modes are of lesser importance because such motions are usually within the ability of the pilot to control, but the short period modes are not, and hence it is required that they be stable. A review of methods for predicting the longitudinal response characteristics of helicopters is given in Ref. 21.

Art. 7–3 was devoted to a brief discussion of the problem of landing impact of aircraft; the important problem of seaplane impact (Refs. 22 and 23) was, however, not mentioned. The water impact of a V-bottom seaplane during step-landing has been analyzed and a generalized theory developed (Ref. 22). The primary flow about the immersed portion of a keeled hull or float is considered to occur in transverse flow planes which may be considered fixed in space and oriented essentially perpendicular to the keel. Because of the absence of a satisfactory three-dimensional theory, the motion of the fluid in each flow-plane element is treated as two-dimensional; the effects of longitudinal components of flow and end losses are accounted for by an aspect ratio type of correction. During the impact, the momentum lost by the seaplane can be considered to be transferred to a portion of the water in contact with the hull, which has a downward velocity equal to that of the body. This mass of water is usually called the *virtual mass*. Since the entire initial momentum of the body is thus assumed to be distributed between the body and the virtual mass, the momentum of the body and the virtual mass are constant throughout the impact. The motion of the body subsequent to the instant of initial contact can then be determined from the simple equation

$$m\dot{y}_0 = (m + m_v)\dot{y} \tag{9–12}$$

where m and m_v are the aircraft mass and the specified virtual mass, respectively, and y_0 is the distance from the keel to the undisturbed water surface in any given flow plane, measured normal to the keel and positive upward. For calculating the behavior of seaplanes during landings in a seaway, the analysis may be applied to rough-water impacts if the initial conditions are defined relative to the wave surface.

More advanced seaplane configurations involving hydro-skis may be studied on the basis of the behavior of a planing surface. Oscillations of such planing surfaces have been encountered (Ref. 24) which were evidently self-excited, rather than forced vibrations resulting from wave action, since they did not depend on the presence of waves on the water surface. The vibration occurs when the wetted length is small compared with the beam and seems to involve the first bending mode. As speed is increased, the first mode is succeeded by an interval of irregular vibration followed by higher mode vibrations. The vibration is often accompanied by a drumming sound which is apparently caused by the rapid succession of impacts of the planing surface with the water surface.

In addition to the problems of landing impact of land and seaplanes, the problems involved in launching aircraft and missiles by catapults (Ref. 25) and rocket boost (Ref. 26) are highly interesting. In the former case, the problem of yawing instability during catapulting has stimulated considerable work. Aircraft are normally catapulted by a towing cable attached to the airplane at a single point (single pendant arrangement) or at two points (V-bridle arrangement). The equations of motion in each case must also account for the effects of tire flexibility and the elasticity of the tow cable. For take-off flight-path analyses under catapult conditions, the complicated landing gear behavior must also be included in the equations of motion because the airplane is usually "snubbed" down at the beginning of the launching with consequent rapid extensions of the shock struts at cable release, causing the airplane literally to hop into the air.

The ground run of aircraft is not without problems and we have already seen (Chapter 4, Prob. 4) the difference in stability characteristics between conventional gear and tricycle gear arrangements. The landing gear itself is subject to several vibration phenomena, the most well-known of which is shimmy (Ref. 27). The oscillation is self-excited and is characterized by a rapid build-up of a high-frequency oscillation, usually involving a rotory oscillation of the wheel about a vertical axis. A complete analysis of the problem involves consideration of the lateral displacement of the wheel while the tire footprint remains fixed on the ground due to lateral flexing of the tire, swiveling motion of the wheel about the strut axis, lateral displacement of the pivot or strut against the elastic restraint of the strut and airframe, relative motion between the swiveling of the wheel and the movement of the damper piston due to the elasticity of the damper linkage, and the relative motion between the strut and the airframe (the airframe itself is represented by a lumped system of one or more degrees of freedom). Neglecting the airframe degrees of freedom, the shimmy problem is seen to involve five degrees of freedom.

Turning now to a somewhat broader aspect of airplane dynamics, we may note that the three primary sources of vibrational energy on a conventional airplane are the engine, the propeller, and aerodynamic flow. Vibration forces arising from engine unbalances and torque fluctuations are important in the design of engine parts and engine mounting structures and constitute important problems within themselves. When the propeller is mounted on the crankshaft it is no longer an individual component, and hence we must consider propeller-crankshaft vibrations. It is true, however, that the propeller has some vibrational frequencies which are not dependent upon its coupling with the engine but which depend upon the flutter characteristics of the blades and on the passage of the blades past some nearby structure such as the wings or fuselage. Finally, the aerodynamic vibrations are considered to be of either a flutter or a buffeting type. In each of these kinds of vibration, there may exist sufficiently high frequencies to create audible sound. For example, consider the propeller, which is one of the most intense and complex sources of noise known. There seem to be various types of noise originating from the propeller, but these can, however, be divided into three classes: rotational noise, vortex noise and flutter noise. The rotational noise consists of a fundamental note and a series of harmonics and is produced by the thrust and by the blade displacing air in both directions perpendicular to the blade. The vortex noise involves a continuous series of frequencies from 1000 to 5000 cycles and is due to the alternate shedding of vortexes from the sides of the blade. The flutter noise is due to elastic vibrations of the blade and hub.

We thus see that an aircraft creates considerable noise covering a very wide range of frequencies. Noise acts as a forcing function, and hence there exists the possibility of structural fatigue in the areas near propeller and jets. The situation has now become so aggravated that noise damage is becoming an important consideration in the determination of aircraft components (Ref. 28 and 29). A fatigue failure of a flat panel, subject to intense acoustic loading, is shown in Fig. 9–6.

As a final example of miscellaneous dynamics problems, let us consider one of the many aspects of aerodynamic heating. If heat is suddenly

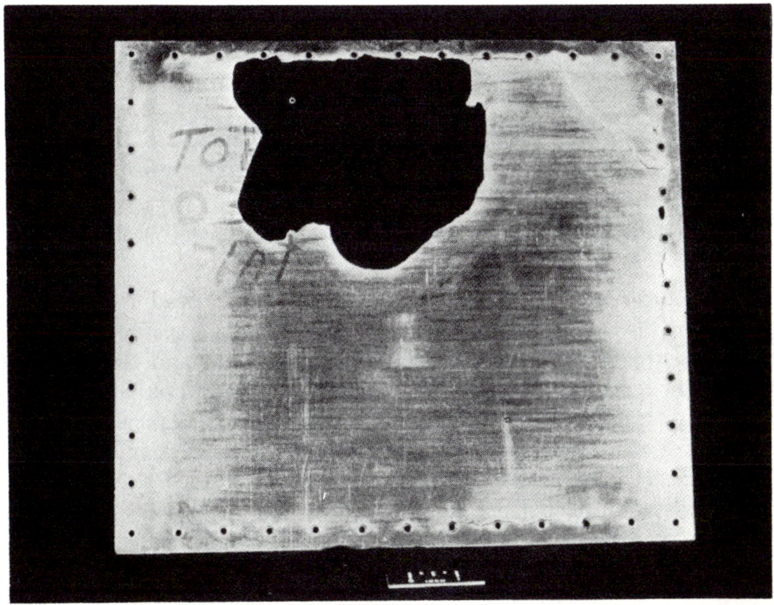

National Advisory Committee for Aeronautics

FIG. 9–6. Failure of thin panel due to intense acoustic loading.

applied to one side of a simply supported rectangular beam, and the effects of inertia of the structure are considered, the beam will oscillate at the lowest natural frequency of the beam about the static axes obtained by neglecting the inertia (Ref. 30). The effect of inertia is of importance in the case of thin plates.

Nonuniform heating over the surface of a wing in high-speed flight results from the effects of aerodynamic friction and causes significant changes in the effective stiffness of the structure. Such changes in stiffness are not the result of changes in material properties but rather depend

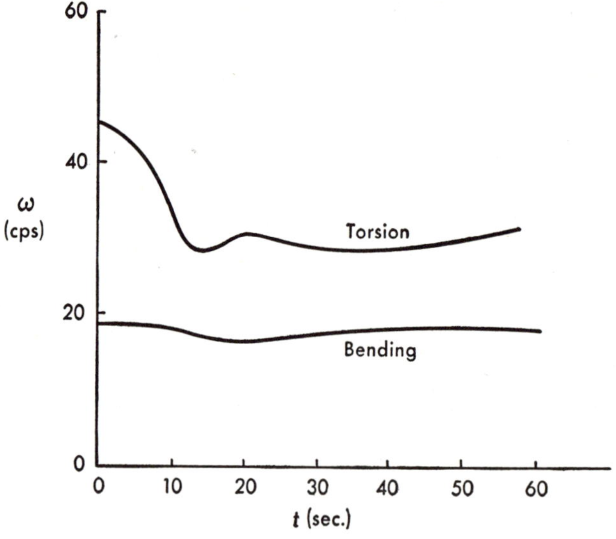

FIG. 9–7. Reduction in natural frequencies as a result of aerodynamic heating.

upon changes in the state of stress and generally lead to significant reductions in the natural frequencies. For example, experiments with a simple cantilever wing model heated at the leading and trailing edges have shown reductions in natural frequencies, as indicated in Fig. 9–7.

Such thermoelastic effects exhibited by structural components promise to become of increasing importance as flight speeds push into the range of significant thermal effects, and the coupling of such thermal effects with aeroelastic phenomena leads to a new class of problems under the heading "aerothermoelasticity."

9–2. Effects of Nonlinearities. Efforts to account for new or obscure phenomena in dynamics problems frequently lead to consideration of various nonlinearities which may exist in the system.

Let us consider for a moment some effects of the introduction of different types of *structural* nonlinearities in the classical flutter problem (Ref. 31). The structure provides elastic restoring force or moment terms in the equations of motion, and hence we may consider three types of nonlinear spring which are characterized by the restoring force or moment as a function of displacement, as shown in Fig. 9–8. The flat-

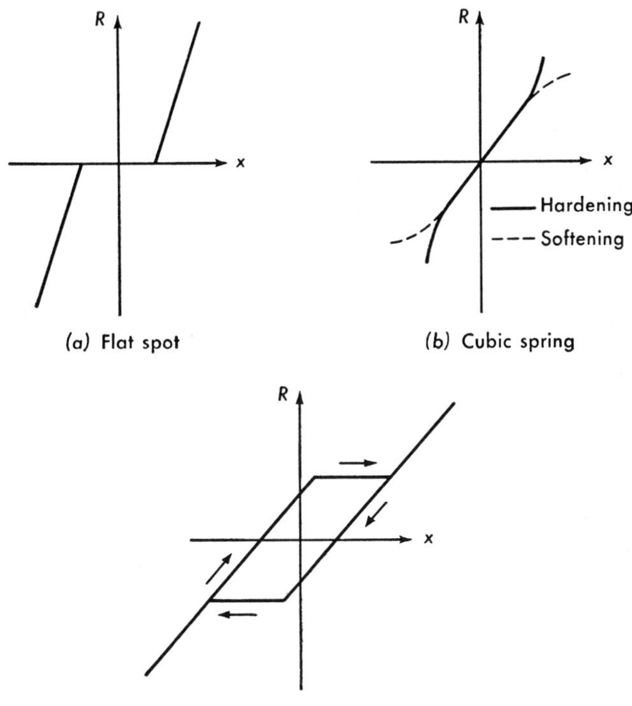

Fig. 9–8. Types of structural nonlinearities.

spot type of nonlinearity may represent free play in a hinge or linkage system of a control system or possible slippage in joints; the cubic spring may be characteristic of power controls or large-amplitude structural deflection; and the hysteresis type of nonlinearity may be characteristic of control surfaces with free play if friction exists in the linkage and also in slippage of joints. It is beyond the scope of this book to present any methods of nonlinear analysis,* but we may consider briefly some of the effects of such nonlinearities.

* For methods of analysis of nonlinear vibrating systems, see McLachlan, N. W. *Ordinary Non-Linear Differential Equations*, 2d. ed. (New York: Oxford University Press, 1956).

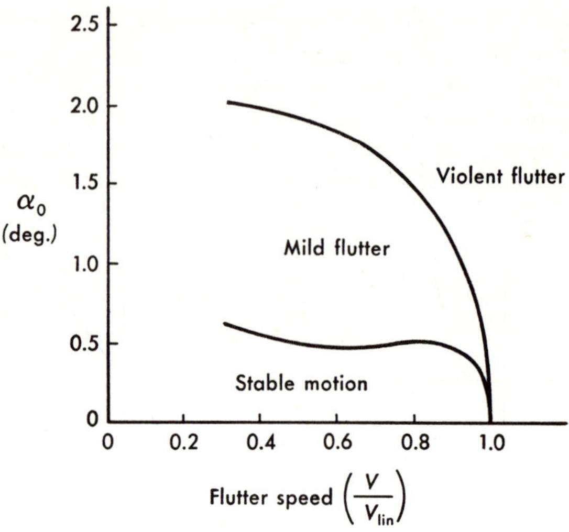

FIG. 9–9. Flutter regions with flat-spot nonlinearity.

One of the principal characteristics of any nonlinear vibrating system is a dependence of amplitude on frequency, i.e., on the amplitude of the initial disturbance. In several of the cases studied (Ref. 31), the flutter speed was decreased as the initial disturbance was increased. For example, the two-degree-of-freedom flutter of a system with free play (Fig. 9–8a) was found to occur well below the flutter speed of the linear system; three

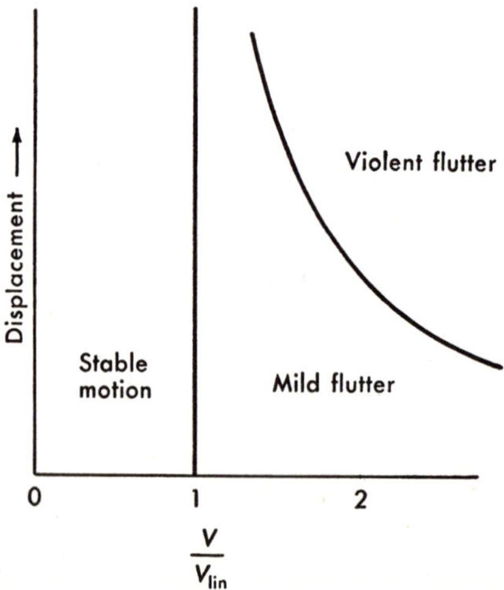

FIG. 9–10. Flutter regions with hysteresis nonlinearity.

distinct regions were found: a stable region, a mild flutter region, and a violent flutter region. These are shown in Fig. 9–9, where the initial angular displacement of the wing is plotted against the ratio of flutter speed to flutter speed of the linear system.

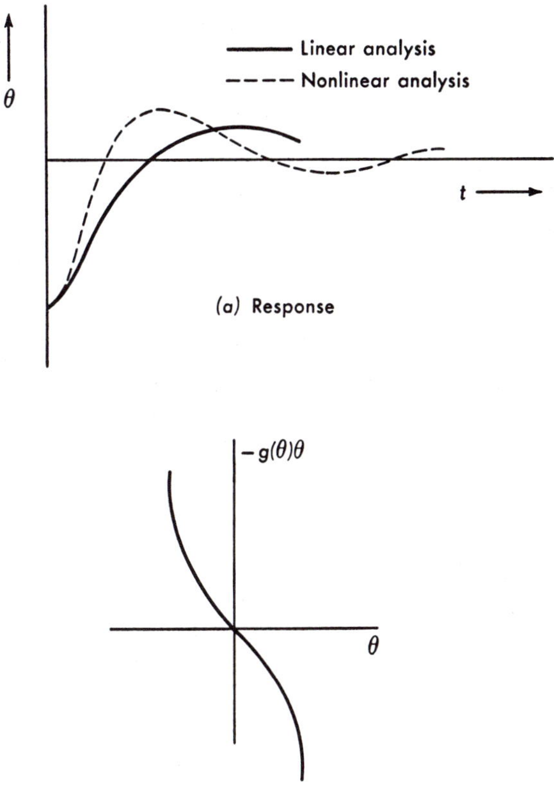

Fig. 9–11. Effect of nonlinearity on flight stability.

A study of flutter with a cubic spring in the system shows that for the hardening spring (solid curve in Fig. 9–8b) the spring is stabilizing, and hence there is little change in flutter speed from that of the linear system. For a softening spring, however, there is a destabilizing effect, with a slight lowering of flutter speed as the initial disturbance is increased.

For the system with hysteresis the three regions of flutter found for the flat-spot case again appear. However, it is found here that only flutter speeds higher than the linear flutter speed are possible. This is so because at small displacements the system oscillates on a linear path through the center of the hysteresis loop, but for larger displacements the damping effects of the hysteresis appear and the free play lowers the stiffness so that a higher flutter speed appears. The interaction of these two effects is also responsible for the region of mild flutter. These flutter regions are shown in Fig. 9–10.

It is also possible that the nonlinearities appear in the aerodynamic terms of the flutter equations. We have already noted that such is the case in stall flutter (Refs. 5 and 6) where the wing is partially or completely stalled. Again it was noted that these different types of flutter may occur in different regions of motion.

Finally, it is possible that the nonlinearity appears in the equations of motion as the result of large amplitude motions of a rigid body. For example, in studying the flight stability of aircraft, the analysis presented in Chapter 8 was limited to small disturbances so that the equations were all linear. Various analyses which include certain nonlinearities of this type, or nonlinearities in the stability derivatives, are given in Refs. 8, 9, 32–35. Fig. 9–11(a) shows the dynamic response to a step-function input in angle of attack calculated by both nonlinear and linear methods (Ref. 33). The nonlinear restoring moment as a function of incremental angle of attack is shown in Fig. 9–11(b). It is seen that the results of the nonlinear and the linear analyses are qualitatively similar, but there is a large difference in the quantitative results.

9–3. Experimental and Analog Methods. Very often it is necessary that experiments be performed to check the validity of various analytical computations made in the dynamic analysis. This is especially true for the natural frequencies and modes and is usually highly desirable for other parameters such as control surface inertias, stiffnesses, and damping coefficients. Experimental confirmation of calculated data is almost mandatory if the airplane is one of unusual configuration so that the analyst cannot rely on past experience, or in other words, if there is any doubt whatsoever as to the reliability of the calculated data. In addition, certain components of the airplane are customarily checked by means of some experimental procedure. For example, the dynamical characteristics of landing-gear systems are almost always checked with computations by means of "drop" tests in which the landing-gear assembly is dropped from varying heights and the pertinent dynamical quantities actually measured.

As a complement to vibration tests of the completed airplane and to the analytical analyses, it is possible to study the flutter and aeroelastic characteristics of an airplane or missile by means of models. These models are so constructed that they are dynamically similar to the actual craft in every way. By placing such a model in a wind tunnel, its dynamical characteristics may be readily observed. This technique has been developed to an extremely fine degree.

In addition to analytical studies, ground vibration tests, and model tests, there exists the possibility of investigating the aeroelastic properties of the airplane in actual flight.

Certain of the elastic characteristics of aircraft structures which are necessary for use in various dynamics problems are measurable in the laboratory. This procedure consists of determining the deflection or deformation characteristics of the structure in the various modes which are expected to prevail in the problem under study. Included in the investigation is the determination of the nature of the elastic deflection under both statically and dynamically applied loads. The elastic axis may, for example, be determined for a simple wing structure by either of two experimental methods, since it is defined as the axis about which the surface rotates under the application of a pure static couple or as the chordwise axis at which a static force will produce only translations of the surface. Following the first definition, a simple chordwise loading arm may be attached to the wing tip and loaded so as to produce a couple, and then the elastic axis can be located by measurement of surface deflections at various points. These known values of applied torque and measured torsional displacements are also useful for measuring the total wing torsional stiffness for use in divergence calculations. Following the second definition, a force may be applied at successive chordwise locations at various spanwise stations and observations made of the leading and trailing edge deflections—the chordwise point of application of the force that results in identical values for the leading- and trailing-edge deflections is the elastic axis point. More recently, it has become the general practice to employ structural influence matrices in analytical work; the elements of the matrix can be determined experimentally quite easily by simple application of loads at the various points on the wing structure, successively, and measuring the deflections at each of the other wing points each time.

Moments and products of inertia of various components, particularly control surfaces, can be determined experimentally in a number of ways. A calibrated torsional pendulum or a compound pendulum is often employed for determining moment of inertia, while product of inertia is often determined by forcing high-frequency oscillations of the surface about one axis and observing the location of nodes resulting in the motion induced about the other axis. Qualitative determination of dynamic balance characteristics of a control surface may be obtained by exciting vibration of the surface, as installed on the airplane, about its hinge line and studying motion induced about other axes.

Often it is of some interest to determine the structural damping characteristics. Because of the low damping of most aircraft structural components, extreme accuracy of measurement of load and deflection is necessary, and consequently the structure must be very rigidly supported so as to eliminate extraneous effects. The loading should be such as to produce a deflection curve approximating that of the mode under consideration. The loading is applied dynamically, and deflections are measured over the surface; the damping coefficient is evaluated from the energy represented by the area enclosed in the elastic hysteresis curve, the spring constant of deflection at the measured point, and the maximum amplitude of cyclic deflection. Damping may also be determined from displacement time curves of free vibration obtained by the sudden release from an initial deflection or by the sudden removal of a vibratory shaker force occurring at a frequency near that of free vibration. The calculation of damping coefficient from the logarithmic decrement has already been discussed in Chapter 1. If the amplitude decay in such a test is linear with time, it may be concluded that the damping is essentially due to dry friction rather than the usual structural and aerodynamic damping. If the displacement curves are obtained from oscillographic records or the output of an electrical pick-up, the first few cycles of oscillation should be discarded because of the presence of various transients.

One of the most important of all parameters is the natural frequency. The two principal methods of determining natural frequencies are those involving measurement of the amplitude of response to a periodic vibratory force and measurement of frequency of free vibration resulting from the application of a step-function force, either by the sudden release of applied static loads or the imposition of impact loads.

A periodic exciting force may be applied to a structure by means of an oscillator or shaker. The shaker may be energized mechanically, pneumatically, or electrically and is arranged so that both frequency and magnitude of the exciting force may be varied over rather wide ranges. Shakers of the mechanical type employ mass unbalance as the source of the applied force and are arranged with a double rotary unbalance, phased so as to produce a periodic force. The amplitude of the structural response may be measured in a variety of ways, ranging from a direct-writing chart moving past the point in the structure at which the response is being measured, to the relative complexity of the electromagnetic pick-up with its attendant power supply, amplifier system, and recording oscillograph.

Detailed information concerning experimental techniques and model studies may be found in Refs. 36–38. Also see Art. 9–4.

Because of the complexity of the analytical computations required in most aeroelastic analyses, there has been developed in recent years a technique for solving the dynamical equations by means of analogous electrical circuits. This procedure is so widely used that a brief discussion of electrical analogs and their application will now be given.

In order to establish the analogy concept, let us consider a simple mechanical system containing a dashpot, c, spring, k, and mass m, and acted on by an external periodic force (Fig. 9–12a). The governing differential equation is

$$m\ddot{x} + c\dot{x} + kx = P \sin \Omega t \tag{9-13}$$

Now let us consider a simple electrical circuit with inductance, L, capacitance, C, and resistance, R, arranged in series with a voltage source (Fig. 9–12b). The governing differential equation is, by application of Kirchoff's voltage law,

$$L\ddot{q} + R\dot{q} + \frac{1}{C} q = E \sin \Omega t \tag{9-14}$$

where the current i is given in terms of the charge q by $i = \dot{q} = dq/dt$. We note that Eqs. (9–13) and (9–14) are *identical in form*, and hence we conclude that an analogy exists between the mechanical and electrical systems. Thus, the series-circuit analogy is obtained by the equivalence between mass and inductance, dash-pot damping and resistance, etc., as

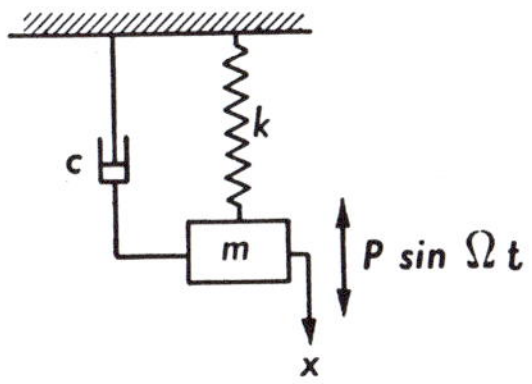

(a) Mechanical system

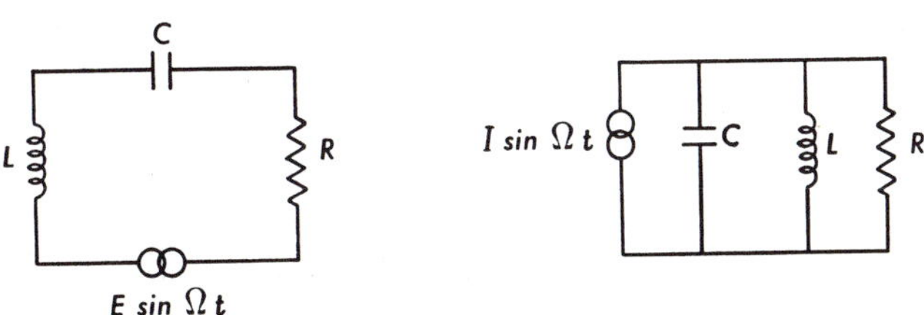

(b) Series circuit (c) Parallel circuit

Fig. 9–12. Electromechanical analogies.

TABLE 9–1

FORCE—VOLTAGE ANALOGY

Mechanical	Electrical
m	L
c	R
k	$1/C$
P	E
x	q
$\dot{x}$	i

shown in Table 9–1. Note that the analogy depends entirely upon the form of the differential equations of the two systems rather than upon any physical similarity.

It is of course quite possible to arrange the elements of the electrical circuit in parallel rather than in series (Fig. 9–12c). In this case the analogy becomes one between force and current, as the differential equation is, by application of Kirchoff's current law,

$$C\frac{dv}{dt} + \frac{v}{R} + \int \frac{v}{L}\,dt = I \sin \Omega t \tag{9–15}$$

where the current i is given in terms of the voltage v by $i = C(dv/dt)$. The complete analogy is given in Table 9–2.

In order to study complex mechanical problems, then, it is often practical to construct the equivalent electrical circuits and then to study the behavior of the circuits. Such analogies have been developed for the study of the vibratory characteristics and stresses and deflections in beams (Refs. 39 and 40). Similarly, complete aeroelastic analyses have been accomplished by use of analogous electrical circuits (Refs. 41 and 42). For ease in doing such work, general-purpose electronic analog computers have been constructed which may be used for a great variety of problems by proper arrangement of circuit elements (Fig. 9–13). Such an

TABLE 9–2

FORCE—CURRENT ANALOGY

Mechanical	Electrical
x	$\int v\,dt$
$\dot{x}$	v
$\ddot{x}$	dv/dt
P	I
k	$1/L$
c	$1/R$
m	C

Reeves Instrument Corp.

Fig. 9–13. REAC® analog computer.

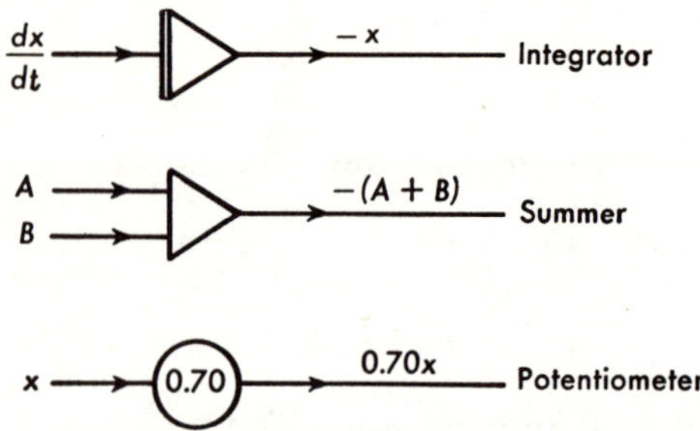

FIG. 9–14. Symbolic forms of analog computer elements.

analog computer involves a system of integrating amplifiers, summing amplifiers, and potentiometers coupled with electromechanical servos for performing nonlinear operations. In addition, special equipment such as limiters, relays, input-output tables, and function generators are used for some operations. The integrators, summers, and potentiometers may be denoted by symbolic forms (Fig. 9–14), and the appropriate wiring diagram determined for the computer. The general procedure for the solution of differential equations on such a computer is to isolate the highest derivative in the equation on the left-hand side and arrange the components so as to solve for this derivative. For example, the symbolic analog circuit for the simple mechanical oscillator governed by Eq. (9–13) is shown in Fig. 9–15. The study of effects of structural nonlinearities on flutter presented in Ref. 31 was accomplished by use of such an analog computer.

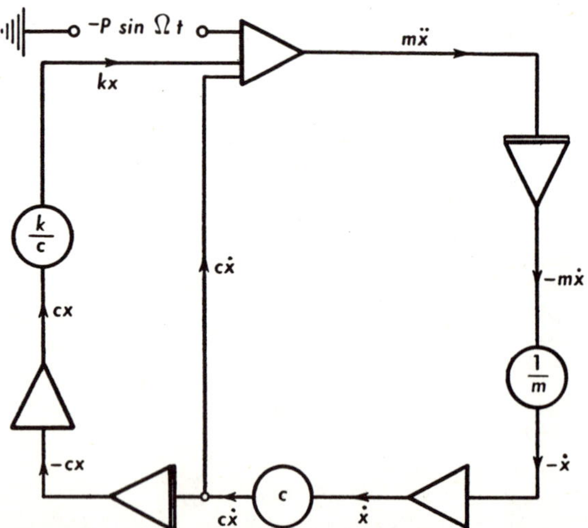

FIG. 9–15. Analog circuit for a simple mechanical oscillator.

A study of the effects of aerodynamic nonlinearities and automatic control on the longitudinal motion of an airplane, by analog techniques, is given in Ref. 35.

Other types of analogies are possible. For example, the equations governing the flow of an incompressible fluid are identical in form with those governing the electrical potential in a shallow tank containing an electrolyte. Such an analogy was used in Ref. 43 to study the aerodynamic aspects of aeroelastic problems.

9–4. A General Survey of Dynamic Aeroelasticity.* The demands of higher speeds and increased structural efficiency have resulted in aircraft of greater relative flexibility. Since airframe distortions are inevitably accompanied by changes in the aerodynamic loading, flexible aircraft display a variety of static and dynamic behaviorisms which require design control, or which must be used to design advantage.

Looking back on the problem of airplane design as it existed some two decades ago, hindsight tells us that a relatively comfortable position had been achieved, at least in some quarters of the engineering department. The aerodynamicist, the structural engineer, and the stability and control specialist could proceed with their respective tasks with a considerable measure of independence. For each design configuration, the aerodynamics group calculated performance factors and flight loads with relative indifference to the structural problem. The structural engineer, in turn, accepted these flight loads and conducted a static analysis to determine strength levels. The stability and control group was content to deal in problems entailing the dynamics of rigid bodies. Dynamic loads arising from aircraft maneuvers were accounted for through a series of simple, statically inspired design criteria.

Over the past decade or so, this neat pattern has been seriously disrupted. In order to calculate the loads encountered during steady flight or during slow maneuvers, the aerodynamicist and the structural engineer find themselves inescapably drawn toward a unified, static aeroelastic approach. No longer able to treat their principal professional areas as independent entities, they are forced to interrelate the magnitudes of the aerodynamic loadings with the structural distortions which accompany them. Stress levels and flight-safety determinations thus become a function of both the aerodynamic and flexibility characteristics of the airframe, the one modifying the participation of the other.

The stability and control specialist likewise finds himself entangled in the aeroelastic net. For aircraft maneuvers involving principally the long-

* This article is adapted from the paper "An Appraisal of Aeroelasticity in Design, with Special Reference to Dynamic Aeroelastic Stability," by Martin Goland, presented at the Sixth Anglo-American Aeronautical Conference, Folkstone, 1957; it is given here by permission of the Royal Aeronautical Society.

period stability modes, he can restrict his analysis to the rigid-body coordinates of the airplane, but the values of the stability derivatives must be modified to account for substantial static aeroelastic effects. For maneuvers involving the short-period modes, in modern designs the mode frequencies are approaching to an increasing extent the lower structural frequencies of the airframe; in such cases, additional aeroelastic degrees of freedom must be introduced in the stability and control analysis.

Perhaps the most dramatic airframe design changes have occurred in the areas of flutter and dynamic loads. The rise in importance of flutter considerations in the modern engineering office has been no less than meteoric. Once occupying a relatively obscure position in the hierarchy, the flutter specialist today finds himself in the forefront of the design team, with flutter margins often dictating the success of the entire design effort.

The flutter engineer, by virtue of his long association with dynamic aeroelasticity, has also been a leader in urging the replacement of statical philosophies for the estimation of dynamic loads by more precise dynamical approaches. Gust loads, rapid maneuver loads, and landing and taxiing loads are today frequently estimated by actual solution of the aeroelastic equations of motion—a final design recognition that aircraft are built to withstand loads applied dynamically, and that the fourth coordinate of the airplane, time, must be dealt with directly in the determination of flight adequacy.

The "aeroelastic triangle" is composed of three professional disciplines —dynamics, structures, and aerodynamics—which enter into the aeroelastic mechanism. With the increasing use of high-response, automatic-control systems which couple dynamically with the aeroelastic modes, there is some reason for changing the aeroelastic triangle into a square, with servomechanism theory forming the fourth corner. For the present purpose, however, we shall use the three-part division as a discussion guide.

Flutter Dynamics. To introduce a survey of flutter dynamics, it is of value to reconstruct briefly the manner in which the subject developed. If we visualize the hypothetical experiment of vibrating an aircraft in a vacuum, and if we presume that the airframe is elastic and undamped, then the vibration problem becomes the much-studied classical one for small vibrations of an elastic continuum; during the vibration, an infinity of undamped normal modes can make their appearance, each uncoupled from all the others, and each distinguished by a characteristic mode shape and frequency.

For the aircraft in flight, an infinity of natural modes are still inherent in the system, although these modes are no longer "normal" in the classical sense. For flight below the minimum flutter speed, all the natural modes are damped and stable. As the minimum flutter speed is approached, one of the natural modes will lose stability; at the flutter speed, the mode will display a zero-damping characteristic. Such a resonance condition is obviously critical in design, since random and even slight disturbances will cause the airframe to vibrate with increasing amplitude until nonlinearities in the system restrict further growth, or until the amplitude becomes sufficiently severe to cause structural failure.

If, for academic reasons, we were to trace the stability of the various natural modes over the entire speed range from zero on up, a spectrum of critical modes might well be uncovered; each mode would gradually lose stability as its flutter speed was approached, generally becoming unstable for slightly higher speeds and then possibly becoming stable once again for still higher flight speeds. First one mode and then another will enter the picture as a "critical" flight mode. For design purposes, of course, it is necessary to isolate and study that mode which yields the lowest flutter speed, and much of the complexity of the flutter problem surrounds the fact that simple methods for accomplishing this are still not available.

Early researchers in the flutter field were quick to observe that classical small-vibration theory, which embraces the resonance concept, was the stepping stone from which the dynamical theory of flutter should proceed. The use of generalized coordinates within the framework of the Lagrange equations of motion, supplemented by the Rayleigh-Ritz methodology, enabled the treatment of the continuous airframe in terms of a finite, small number of degrees of freedom. It soon became clear that the generalized coordinates should preferably consist of the normal modes of the airframe, or of significant airframe components such as the wing and tail. Guided by the theorems of the classical theory, researchers concluded that a particular flutter mode would be influenced significantly only by those normal modes with natural frequencies in the vicinity of the flutter frequency. The flutter problem was thus reduced to a form capable of being handled in the design office, a matter of especial importance during the years when computing services were restricted to the slide rule and the desk calculator.

Unfortunately, however, the early impetus given to flutter theory by the classical discipline of small vibrations was not sustained by the appearance of an expanding body of fundamental knowledge of the dynamical mechanisms of flutter. The virility of the classical theory arises from a fortuitous combination of physical and mathematical circumstances; in particular, the vibrations of an undamped system can be described completely by a kinetic and a potential energy, each being a quadratic form in the system velocities and displacements, respectively. This elegant simplification is absent in the flutter problem and accounts for the fact that both physical and mathematical interpretations of the flutter mechanism are elusive and, at best, difficult to follow.

With fundamental progress delayed, experience and judgment became the mainstays of successful flutter design. Even today, flutter analysis remains very much an art. In any new model aircraft, a diversity of flutter modes requires examination. The flutter specialist must decide which of these warrants concentrated (and lengthy) study to determine its design criticalness, and he must also determine which normal modes should be used as degrees of freedom in describing the flutter mode behavior.

The true artist, it is often said, concentrates on the art of omission. The success of the flutter engineer is likewise dependent on how much he can, for good reason, delete from his analysis. The investigation of a non-critical flutter mode or the inclusion of an unnecessary degree of freedom represents wasted design time and money. The sword hangs by a thin thread, however, and the penalties of improper omission may be even more severe. A casual investigation which does not bring the minimum flutter speed to light may lead directly to a design catastrophe.

Although fundamental theorems which can make flutter design more rational and precise have been slow in making their appearance, this must be attributed more to the difficulty of the problem than to a lack of interest on the part of researchers. In his excellent general text on aeroelasticity, Fung (Ref. 44) briefly discusses the work of Flax, Lanczos, Wielandt, and others, relating to the nature of "non-self-adjoint" equations, this being the type of mathematical formulation often encountered in static aeroelasticity and underlying the whole of dynamic aeroelasticity. (The classical vibrations of conservative systems are described by "self-adjoint" equations, which are typified by symmetry of the structural and inertial matrices; in non-self-adjoint systems, the determinant describing the eigenvalues is unsymmetric.) In addition to the references given by Fung, mention should be made of work by Serbin (Ref. 45), in which an attempt is made to clarify the modal concepts pertinent to flutter systems.

Some years ago, the bending-torsion flutter of a simple wing was studied (Ref. 46) by a technique which is structurally and dynamically exact, although aerodynamically approximate. Further studies along the same lines were later carried out by staff members of the National Advisory Committee for Aeronautics (NACA). Although the method does provide an accurate description of the flutter mode dynamics for systems composed of "beam-like" elements, its real utility is masked by the fact that general conclusions are not reached, and calculations for specific systems are unduly laborious.

As an illustration of one of the pressing problems in flutter dynamics requiring practical clarification, consider the question of "hard" versus "soft" flutter. The former is catastrophic in character, i.e., the system remains heavily damped at all speeds practically up to the flutter speed; within a very narrow speed range below the critical value, stability is rapidly lost and, at the critical speed, flutter occurs with explosive violence. Wing bending-torsion modes are usually of this catastrophic nature; in flight, practically no warning is given of the approach to flutter, and structural failure is almost certain when the aircraft enters the critical region.

"Soft" flutter, on the other hand, causes the airframe to vibrate with undesirable, but self-limiting, amplitudes. Such modes, which are usually associated with control surface and tab flutter, cannot be tolerated in the final design, but when first encountered will not result in severe damage to the aircraft if the pilot is alert and skillful in taking counteractive measures.

At present, only approximate criteria are available for determining, in advance of model or flight tests, the extent to which a particular flutter mode is explosive. Nevertheless, there are several practical reasons which make it desirable to have this information. Hard flutter modes must obviously be approached with considerable design conservatism; with softer modes, on the other hand, more liberties can be taken to approach complete system efficiency, relying on subsequent experimentation for design refinement. The entire question of safety during flight-flutter tests is also vitally dependent on how well the engineer can appraise the violence of the modes which will be encountered in flight. Even when designing flutter models, it is valuable to estimate the character of the modal behavior, since elaborate flutter "brakes" must be employed with hard modes to avoid destruction of the model.

The violence of a flutter mode is dependent on two general characteristics of the system—its behavior in the range of small (linearized) oscillations, and its behavior when undergoing larger-amplitude, nonlinear vibrations. It is generally thought that the performance of the system in the linear range can be appraised from conventional flutter calculations; a sudden and drastic dwindling of the system stability, starting just below the flutter speed and continuing up to the flutter point, is believed to indicate the presence of a violent mode. The large-amplitude, nonlinear oscillations of flutter systems are at present understood only in barest outline, principally due to the difficulties of interpreting both the dynamic and aerodynamic mechanisms.

As a final item in this discussion of flutter dynamics, mention should be made of the ground-vibration problem. As already noted, it is usual in flutter analyses to employ the normal modes of the airframe as degrees of freedom. The theoretical value of this approach is accompanied by several experimental advantages. To begin with, the calculated mode shapes and frequencies can be compared with the ground resonance modes observed during vibration tests of the full-scale prototype airplane or of a suitable airframe model. These ground resonance modes, if pure, will closely approximate the frequencies and shapes of the true normal modes. Such a comparison provides an over-all check on the validity of the inertial and structural data used in the analysis. In some instances, it is desirable to use the measured ground-mode data directly in the flutter calculations, thus eliminating the need for the normal mode computations.

Even when the flutter analysis is formulated on an influence-coefficient basis (more will be said of this later), it is usual first to calculate the normal modes in order to have the advantage of the check with experimental ground-mode observations.

The difficulties of exciting pure resonance modes during ground-vibration tests are by no means slight, particularly for structurally complex airframes. A variety of experimental techniques have been proposed, some of an elaborate character and employing a number of harmonic vibration-exciting units whose frequencies and relative phases are centrally controlled. If better ground-vibration techniques are to be developed, a more precise theoretical understanding must be reached of the vibration behavior of complex aircraft structures. In particular, the manner in which a damped structure can be best excited to yield the normal mode shapes must be further explored. An attempt in this direction has been made by Fraeijs de Veubeke (Ref. 47).

The calculation of the normal modes for a complicated airframe is also a task of formidable proportions. Since, however, the principal difficulties arise in the analytical treatment of the structural behavior, discussion of this is deferred to a later portion of this article where structural problems are dealt with more specifically.

Flutter Aerodynamics. In recent years, a substantial amount of both research and design interest has been shown in matters relating to unsteady aerodynamics and flutter airforces. Progress has been impressive, in the large, particularly when one considers the suddenness with which the aircraft industry has been forced to extend its sights from the subsonic regime to the entire compressibility spectrum. As a consequence of this concentrated attention, a number of excellent surveys of the subject have appeared in print. Garrick (Ref. 48), for example, recently reviewed current theoretical progress in a paper which is remarkable for its concise yet lucid view of the main features of the entire area of flutter aerodynamics.

In the earlier years of flutter study, the aerodynamics problem was considerably simplified by the fact that wings and tail surfaces were "beamlike" structures of reasonable-to-large aspect ratio. The flutter motions at a spanwise station of such surfaces could be specified in terms of a vertical translation and pitching motion of the airfoil profile, plus a control surface rotation where such elements are present; chordwise (camber) deformations could be safely ignored. The aerodynamics problem was thus restricted to the consideration of arbitrary spanwise variations of the three coordinates defining the instantaneous position in space of a station profile.

Flutter commenced as a practical art, to all intents and purposes, when the theoretical solution became available for the flow of an incompressible, inviscid fluid past a two-dimensional (infinite aspect ratio) wing of vanishing thickness, performing small translatory, pitching, and control surface motions. Valuable design results were obtained using the two-dimensional aerodynamic coefficients, uncorrected for the effects of finite span, it being appreciated that such a procedure invariably led to conservative design predictions. Compressibility effects, which gradually grew in importance as high subsonic speeds were achieved, were handled in rudimentary, approximate fashion.

In subsequent years, research on the two-dimensional potential-flow problem continued unabated, until we have now achieved a reasonable theoretical understanding of both the subsonic and supersonic aerodynamics of the thin wing undergoing small oscillatory motions both as a rigid profile and with arbitrary camberlike modes of deformation. The transonic problem, unfortunately, still remains obscure. It is interesting to note that while thickness effects are not found to be of importance when calculating subsonic flutter airforces, they cannot be ignored in many supersonic analyses.

The effects of finite aspect ratio for the low-speed wing were first investigated mathematically by Cicala, who extended the Prandtl-Lanchester lifting-line theory to the oscillating-airfoil case. The use of the Cicala theory in flutter calculations, however, did not prove successful. When comparing the calculated critical speeds with those measured with wind-tunnel models, the calculations were frequently found to be on the unconservative side; this immediately mitigated against the use of the Cicala theory for design purposes. While a variety of theoretical procedures have been advanced for unswept and swept planforms as improvements of the Cicala approach, many of them arrived at by elegant reasoning, all appear to have the same vital failing—one cannot be certain that the computed flutter speed will meet the required engineering accuracy, and the possibility exists that there may be an unconservative prediction.

It is on this account that many flutter engineers even today use two-dimensional "strip" theory to calculate flutter airforces. For refined design, a finite-span theory is used for a second computation, with judgment and model experiments often fixing the final design decision.

A significant recent trend in flutter aerodynamics arose from the need to deal with the low-aspect-ratio wing. For such wings, the flutter mode shapes are "platelike," rather than beamlike, and important camber deformations are present. Thus, the simplification of dealing with a rigid, flapped profile is no longer valid. Instead, an acceptable method for aerodynamic analysis must be capable of treating wings of arbitrary planform, vibrating in an arbitrary, platelike modal pattern. While several alternate approaches have been suggested for the aerodynamic analysis of low-aspect-ratio wings, the most practical design techniques will probably be derived from the formulation of the problem in terms of the *influence coefficient* concept. One such formulation entails the use of the pressure-influence coefficient and can be described in general terms as follows:

The wing planform is divided into a number of small, adjoining areas, each area becoming an aerodynamic "box." For simplicity, let it be assumed that each box is sufficiently small to permit the assumption that the aerodynamic pressure loading over it is uniform, and that the wing downwash over each box is likewise nearly constant. (As in most procedures of this kind, the theoretical formulation of the problem becomes exact when the area of each box is made vanishingly small, the number of boxes needed to cover the wing then becoming infinite. It should also be noted that the assumption of uniform conditions over each box can be improved, if necessary, through the use of "weighting" functions which permit linear or higher-order variations of pressure and downwash over the box area.)

Now consider the hypothetical circumstance when one box is pulsated at a given frequency with a unit downwash amplitude, all other boxes remaining stationary. Such a hypothetical wing motion will result in a uniquely determined pressure distribution over the wing area. The pressure-influence coefficient P_{ij} is then defined as the pressure acting at an arbitrary box j due to the unit downwash pulsation at box i. The calculation of the P_{ij} for various Mach numbers, pulsation frequencies, and for various locations of the boxes i and j over the wing planform is a fundamental problem in aerodynamic theory, and the results can be presented in convenient, tabular form. The aerodynamic loads calculation for a specific wing in flight and a specific vibrational pattern can now be formulated in the following manner: The downwash w_j acting at each box is known from the flight speed, modal pattern, and vibrational frequency. At a particular box i, the total aerodynamic pressure is then $p_i = \sum P_{ij}w_j$, where the summation extends over all the wing boxes.

An alternate formulation of the problem entails the use of the downwash influence coefficient, which relates the downwash induced at box j due to a unit pressure pulsation at box i. This approach evidently bears an inverse relationship to the former one employing the pressure-influence coefficient.

This brief outline of influence-coefficient methodology obviously avoids the theoretical detail required to evaluate the unit-effect coefficients. Over the past several years, various scientific teams have been quite successful in arriving at theoretical expressions useful for tabulating the influence coefficients for wings in subsonic flight and for wings in supersonic flight with all-supersonic edges (the components of the stream velocity normal to the wing edges are supersonic). For wings in supersonic flight but with subsonic edges, continuing research is required in order to arrive at suitable practical procedures. Even with the theoretical difficulties overcome, the numerical labor associated with tabulating the influence-coefficient values is by no means inconsiderable, but this is now being accomplished through the use of high-speed computing machinery.

The conclusion that the influence-coefficient approach holds the key to future design practice can be based on several supporting arguments. To begin with, the methods can be developed so as to be applicable to wings of arbitrary planform and vibrational characteristics, an advantage of considerable significance in this age of changing and novel aircraft configurations Also, despite the complicated theoretical background of the methods, their application in design can be approached on a semiroutine basis. This has obvious merit in view of the current and anticipated short supply of experienced flutter specialists. Finally, as will become clear from later discussion, the most versatile and fruitful formulation of the aeroelastic problem for low-aspect-ratio wings involves the use of influence-coefficient techniques to describe the structural, as well as the aerodynamic phase of the over-all mechanism. Design methods based on this formulation are versatile and powerful, and also possess certain advantages (to be described later) from the viewpoint of the structural engineer.

The discussion of flutter aerodynamics has thus far remained within the province of theoretical design methods. It is now pertinent to review the area of experimental aerodynamics, both as a topic in its own right and to appraise to what extent theory and experiment are in accord. It is unfortunate that some of the most significant research in flutter aerodynamics has not been released for general distribution, so that progress can be traced only in broad outline.

In steady-state aerodynamics, the trained observer cannot fail to be impressed by the advanced state of the experimental art, and the manner in which theory and experiment are interwoven into a powerful, unified whole. By contrast, flutter calculations in the modern design office seldom take direct account of experimental aerodynamic measurements. Even in the research area, careful coordination of the theoretical and experimental branches is all too infrequent. To appreciate the reasons behind this state of affairs, consider the very substantial difficulties which stand in the way of obtaining accurate measurements of flutter airforces. Three kinds of model study are in common use for this purpose—the rigid, oscillating model; the elastic model; and the flutter model. Each will now be discussed briefly.

Rigid, oscillating models are used for the experimental study of the two-dimensional airforces on chordwise-rigid, flapped profiles. Such models also find limited use in measurements of the airforces on finite-span wings, although the character of the wing motion is then somewhat academic, since the wing is not deforming in an elastic modal pattern. The construction and testing of models of this type pose straightforward, although not easily overcome, engineering problems. The mechanism by which the wing is mounted in the wind tunnel must enable the model to execute translatory, pitching, and flap oscillations over a substantial frequency range. Moreover, careful dynamic design is required to ensure that the magnitudes of the flutter airforces are not unduly small compared with the inertial, elastic, and extraneous forces present in the system; if this is the case, accuracy cannot be achieved in measuring the airforces. It is evident that the solution to these problems becomes increasingly difficult as the test Mach number is raised. The conduct of the wind-tunnel tests is also lengthy and expensive, in view of the need for numerous test runs at various combinations of speed and frequency.

The measurement of the airforces on elastic models permits the study of realistic wing modal patterns but raises engineering difficulties of truly formidable proportions. To begin with, the model must be sufficiently light in weight and flexible so that the magnitudes of the air forces are not completely masked by the inertial and elastic forces. Also, interest is usually centered on the spanwise (and, in some cases, chordwise) distribution of the airforces; the instrumentation problems involved in making such measurements are of the greatest complexity. The model must be capable of oscillating over a range of frequencies, unless a large number of expensive models are to be used, and the effects of frequency on the modal pattern must be controlled. Moreover, various modal patterns should be studied, and this will almost certainly entail the construction and testing of numerous models. Finally, as the test Mach number increases, the practical problems of model design become even more severe. Faced with this array of engineering difficulties, and with an associated financial burden of considerable magnitude, it is small wonder that flutter researchers have tended to avoid this experimental path.

The most popular approach to experimental studies in flutter aerodynamics has been through the medium of flutter models (see Figs. 9–16 and 9–17). In this technique, a flutter model is constructed and its critical speed and frequency are measured. If a theoretical flutter calculation for the model is then carried out, using a particular aerodynamic theory, the results of calculation and experiment can be compared. If the critical speeds and frequencies are in agreement, the correctness of the aerodynamic theory is presumed. (This procedure implies that the structural and inertial characteristics of the model are known accurately.

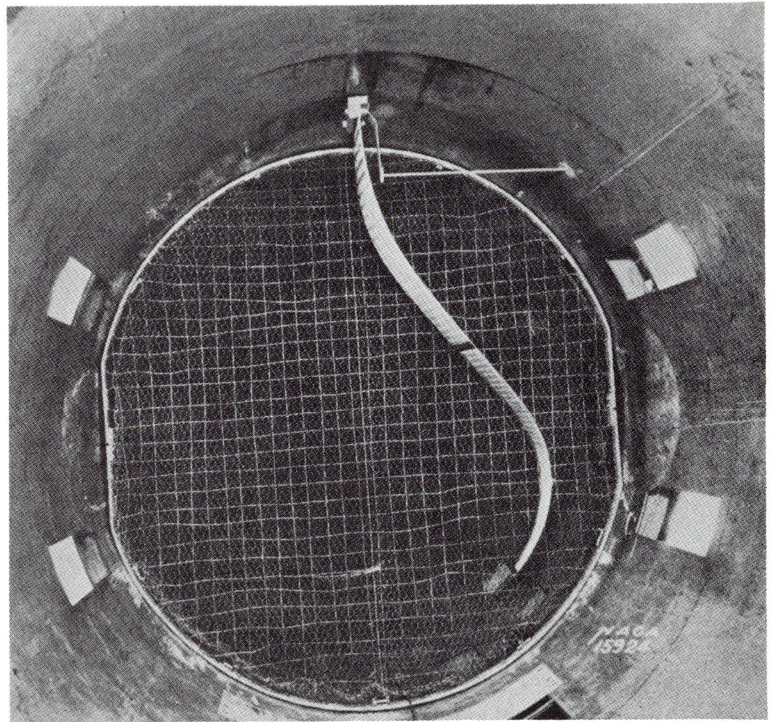

National Advisory Committee for Aeronautics

Fig. 9–16. Wing model in violent flutter.

This knowledge can be gained by studying the still-air vibration modes of the model, by measuring the deflection characteristics of the model under known, static loadings, and by other analogous procedures.)

A number of shortcomings to this approach at once suggest themselves. The evaluation of the merits of the theoretical aerodynamics is based on inductive reasoning, rather than on a direct comparison of the calculated and observed airforces. The construction of flutter models and the associated calculations are an expensive and time-consuming project. Moreover, in order to cover a range of possible flutter configurations, large

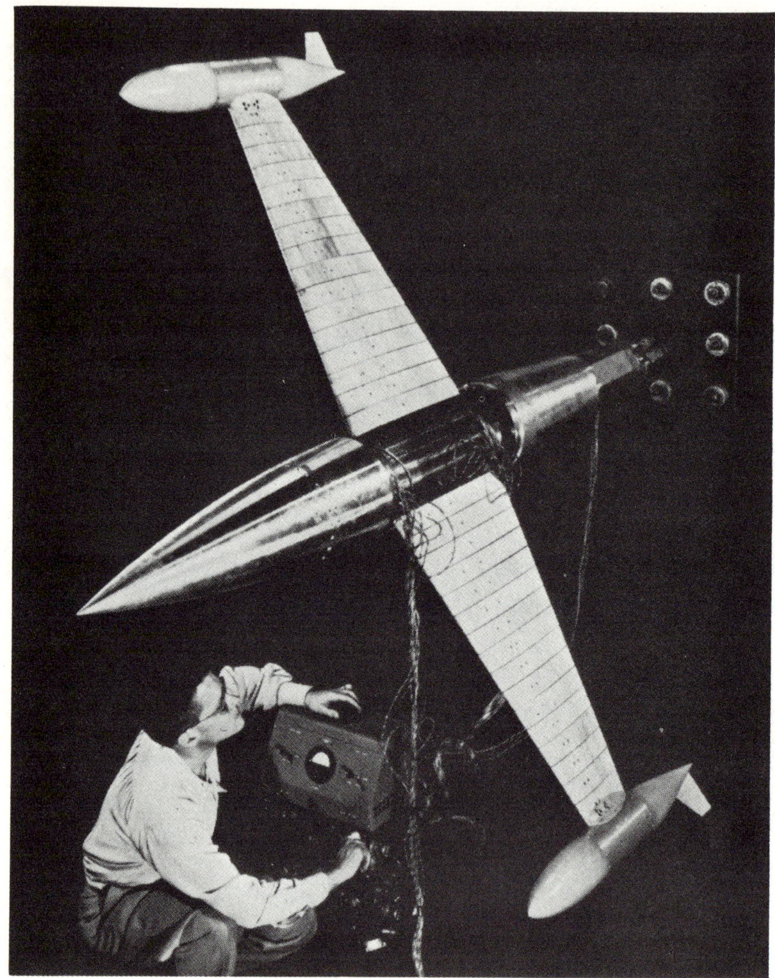

National Advisory Committee for Aeronautics

FIG. 9–17. Complete flutter model.

numbers of models must be used. These disadvantages are to some extent counterbalanced by the fact that flutter data obtained with models of this type can be used to serve the dual purpose of collecting flutter trend data, while at the same time providing information for aerodynamic research.

A survey of the published material on experimental flutter aerodynamics leads to the following general thoughts: First, the available data are sparse and largely incomplete, particularly in the higher Mach number ranges. Reynolds' number effects and boundary layer influences, in particular, are not well understood, even though the importance of these factors has been demonstrated in several test series.

The status of experimental studies is well typified by the situation with regard to two-dimensional airforce coefficients. For rigid translatory and pitching motions of an airfoil in incompressible flow, measurements over limited parameter ranges have shown reasonable agreement with thin airfoil theory. (The magnitudes of the measured coefficients fall slightly below the theoretical values, just as they do in the steady-state case.) For flap motions, the agreement is by no means entirely satisfactory. None of the measurements, however, has made a sufficient impact on flutter design practice to effect replacement of theoretical coefficients by experimental values.

For airfoils of finite span at low Mach numbers, limited measurements show that the several finite-span aerodynamic theories properly predict the coefficient trends for varying aspect ratio and reduced frequency. The precision and range of the experiments, however, have not been adequate to permit a definitive rating of the merits of one theory versus another. A general conclusion is that none of the presently available finite-span theories is entirely adequate for design purposes.

The area of nonpotential flow mechanisms includes the stall flutter problem, boundary layer considerations, and questions relating to the accuracy of the Kutta (smooth off-flow) condition at the trailing edge of lifting surfaces. Stall flutter refers to the dynamic instability of wings operating at sufficiently high incidence so that the range of linear aerodynamics is exceeded. The presently available studies of this phenomenon are largely empirical in character.

Boundary layer considerations become of extreme importance when the transonic flow regime is entered. While such behaviorisms as control-surface "buzz" are probably a consequence of boundary-layer mechanisms, current knowledge of the subject is sadly lacking.

Structural Considerations. The early years of aeronautical engineering were favored by the ability to draw heavily on the past in one principal area, that of structures. Theoretical and practical experience in civil and marine engineering, transplanted to airframe design, served as a sound basis for rapid progress in the new field. For conventional, high-aspect-ratio configurations, structural design techniques were developed to permit the synthesis of airframes of truly remarkable efficiency.

Within recent years, however, several new problems have arisen to challenge the capability of the structural engineer to meet his design responsibilities. For example, the classical basis for wing design has been to view the structure essentially as a bent and twisted beam; for the low-aspect-ratio, platelike wing this approach must be severely modified or completely discarded, and the structure must be dealt with as a redundant assembly of beams and plates extending both spanwise and chordwise. The advent of high Mach number flight has introduced the problem of aerodynamic heating and its counterpart of thermal stresses.

The demands of aeroelasticity have brought into focus still another significant change in the perspective of the structural analyst. In the past, the principal objective of the structural engineer was to estimate the magnitudes of the flight stresses to ensure that overstressing did not occur. Today, however, he must be equally concerned with evaluating the *deflection* characteristics of the airframe, since the aeroelastic specialist must be supplied with a precise picture of how the structure deforms under dynamic loads.

This transfer from an almost complete preoccupation with stress levels to the broader viewpoint of studying both stresses and deflections entails more than simply a change in design-office routine. Analytical and experimental techniques which are adequate for checking flight stresses will not necessarily afford the accuracy required for evaluating deflection characteristics. An illuminating study in this regard has been reported by Hedgepeth (Ref. 49). Even in the case of simple box-beams, his results show that secondary structural effects (such as the effects of shear lag and transverse shear) must be taken into account if the vibrational characteristics of the beam are to be adequately analyzed. This is true even for long beams when the higher modes are considered.

In general, it can be concluded that a pressing demand exists for further structural research on both low- and high-aspect-ratio airframe configurations. For the aeroelastic treatment of platelike wings, the structural engineer is today forced to depend heavily on experimental determinations of structural performance. For modern high-aspect-ratio aircraft, higher modes are becoming increasingly important in fixing the critical flutter speed; here, too, current methods of structural analysis do not live up to the design requirements.

As in the case of flutter aerodynamics, increasing interest is being shown in the use of influence-coefficient techniques for the formulation of the structural phase of the aeroelastic problem. The coefficient most commonly defined for structural purposes is the *deflection-influence* coefficient, which relates the deflection observed at some arbitrary box j to the application of a unit transverse load at a box i. Considerable research attention is being devoted to the development of suitable methods for determining these coefficients for complex structural configurations.

It is evident that one advantage of this approach resides in the fact that deflection-influence coefficients are ideally suited for experimental determination through static test of either the prototype airframe or a structural model. While certain practical test difficulties do arise, they are of the type which will be readily overcome as field experience is gained. The test results can then be used for direct comparison and verification of calculated values or, when necessary, the experimental coefficients can be introduced directly into the natural mode and flutter speed computations.

Along the latter lines, it is of interest to note that the use of influence coefficients has significant implications in regard to the analytical methods employed in aeroelasticity. Thus, the flutter equations of motion for an aeroelastic system can be stated in the following compact (and elusively simple) form: If the system is divided into a series of "boxes," then the deflection z_i at box i is given by

$$z_i = \sum_j D_{ij}(m_j\ddot{z}_j + L_j) \tag{9–16}$$

where D_{ij} is a structural deflection-influence coefficient; m_j is the mass of box j; $m\ddot{z}_j$ is the instantaneous acceleration at box j; L_j is the instantaneous aerodynamic loading on box j; and the summation extends over all the system boxes. Note that the product $m_j\ddot{z}_j$ is the inertial loading on box j; when added to the aerodynamic load, the total load acting on the structure at box j results. An equation analogous to Eq. (9–16) is written for each box of the system.

Now, if the aerodynamics are linear, then for a given flight condition and vibrational frequency, the aerodynamic loading L_j can be related to the instantaneous values of the various deflections z_i. To do this, aerodynamic pressure-influence coefficients are used in the manner indicated earlier during the discussion of flutter aerodynamics, and the downwash w_i at each box is written in terms of the deflections z_i and their time derivatives.

For a system with n boxes, Eq. (9–16) is thus converted into a system of n homogeneous equations in the n unknowns z_i. These equations lead to an eigenvalue problem identical in form to that encountered in all mathematical flutter solutions, and the flutter condition can be determined by the conventional techniques. Note that if the aerodynamic terms L_j are deleted from the equations, the remaining mathematical system defines the normal modes of the airframe.

The significant point raised by Eq. (9–16) is that the problem formulation is in terms of the degrees-of-freedom z_i. This is in contrast to the more usual approach of using the normal vibrational modes of the airframe as the analytical degrees of freedom.

As aeroelastic systems become more complex, the value of influence-coefficient approaches is enhanced. Hence, it is to be expected that flutter analysts will be attracted to problem formulations of this type. Much research remains to be done in evaluating the design-office merits of this approach, as compared to the modal concepts now in common use.

At least two additional topics require mention in this brief survey of structural aeroelasticity. The first is the matter of evaluating the internal damping which is present in the vibrating airframe. Structural damping arises from frictional rubbing at the joints and connections which bind the airframe assembly together, as well as from the internal damping of the materials of construction. Other sources of internal energy dissipation may also be present; the sloshing of fuel in partially filled tanks is one such example. Since flutter motions are induced when the energy dissipated within the aeroelastic system is balanced by the energy extracted aerodynamically from the airstream, it is evident that in many circumstances the magnitude of the internal damping will exert important control on the critical speed.

For flutter analysis purposes, the internal damping in an aeroelastic system is usually described in terms of an empirical *structural-damping coefficient*. In essence, the dissipation mechanism is assumed to arise from damping forces which are proportional to the elastic deflections of the system but which are insensitive to the vibrational frequency. During ground-vibration tests of the prototype structure, decay measurements (or their equivalent) are made for each vibrational mode, and these data serve as the basis for evaluating the associated structural-damping coefficient.

Only casual study is required to show that this description of the energy-dissipation mechanism in aircraft structures is at best a crude approximation of the true circumstances. A theoretical impasse is reached, for example, when one attempts to generalize the statement of energy dissipation to account for transient deformations. In dynamic loads studies, where internal damping affects the extent of dynamic over-shoot, a physically valid description of the damping mechanism is not now available. Moreover, the current use of an empirical damping coefficient prevents the effective extrapolation of past experience to new forms of construction.

Flutter Models and Flutter Testing. The flutter model plays an extremely important role in aeroelastic design. The discussion thus far has amply indicated that many areas of uncertainty must be accepted when undertaking flutter calculations for high-performance aircraft. When the calculations are supported and refined by dynamic model tests, the dependability of the design process can be significantly improved.

Flutter models generally fall into two categories: those used for research purposes, and those employed as aids to the aircraft designer. Since the purposes of a research study can usually be met even when the model has a relatively simple and idealized structure, the principal difficulties encountered in the design and construction of research models relate to problems of special instrumentation and to the requirement that the model be capable of convenient modification so that the widest possible range of flutter parameters can be covered. In the case of high-speed models, considerable ingenuity must be exercised in order to devise structures with adequate strength but with the requisite flexibility and lightness to permit simulation of practical aircraft parameters (see Fig. 9–17).

The problems of synthesizing flutter models for use in design are of a somewhat different variety. Various reasons make it impractical to construct the model as a geometrical scaling of the prototype airframe; chief among these is the relatively modest size of flutter models, which is dictated by the dimensional capabilities of the wind tunnel or other facility to be used for the model tests. The flutter model must therefore be conceived as a mechanical analog of the full-scale system, identical in its aeroelastic characteristics but not necessarily resembling the prototype in structural detail.

The scaling considerations for a flutter model reduce essentially to ensuring that the model and prototype have the same exterior geometry and the same still-air vibrational characteristics, at least insofar as the important flutter degrees of freedom are concerned. Moreover, the density of the model must bear a correctly scaled relationship to the density of the test airstream. It is also evident that the Mach and Reynolds' numbers of the test airstream must be similar to those for full-scale flight. Since it is almost never possible to scale the model correctly in all respects, it is clear that the results of flutter model tests must always be judiciously interpreted, usually from the perspective of a set of flutter calculations.

The difficulties of achieving the proper vibrational characteristics in the model, particularly when higher modes are involved, should not be underestimated. Several useful techniques can be employed which permit the accurate scaling of particular modes, but these are not always successful in maintaining an authentic relationship over the entire modal spectrum. Suitable simulations of control systems also pose problems, and the scaling of such factors as internal damping sources is not readily accomplished. It is clear that the model tests will not be entirely accurate unless complete similitude between model and prototype is achieved; in extreme cases, the model and prototype may flutter in different modes so that the model studies will be entirely misleading.

In general, it would appear that the use of flutter models provides no panacea for the designer's problems. Model programs are both lengthy and expensive and hinge on the availability of large wind tunnels or analogous test facilities. The flexibility of the model approach in keeping pace with design changes also leaves something to be desired. There has been a slight trend toward principal reliance on flutter-model testing for design purposes. It would seem that this is an unsatisfactory design policy. When model tests and flutter calculations are meshed so that areas of uncertainty in the one technique are balanced by the other, the optimum approach to accurate design will be realized.

It has become generally accepted within the industry that the structural design of an aircraft should proceed on the basis of calculations, supported by static test of the complete airframe or of selected components, the process being concluded by flight testing of the prototype as a final demonstration of flight safety. It is not surprising, therefore, to find that the thinking of flutter specialists runs along parallel channels and that full-scale flight-flutter testing is often recommended as the final step in verifying the aeroelastic safety of a new aircraft.

Despite the appeal of full-scale flight-flutter testing, most design groups are at present reluctant to engage in this type of test because of its potential danger. Limited experience collected thus far indicates that flight-flutter tests can be carried out with comparative safety when the critical mode is of the soft variety; the oscillation amplitudes appear to be self-limiting, although not always of sufficiently small magnitude to prevent partial damage should the flutter speed be inadvertently reached or exceeded. However, at least one classic accident serves as convincing warning that when a hard flutter mode is approached, current flight-test techniques are not adequate to ensure that catastrophic failure will be avoided.

The usual test approach is to excite the airframe into vibration at gradually increasing speeds and to measure the aeroelastic stability at each speed over the pertinent frequency range. The approach to flutter is detectable by the loss in stability which precedes the flutter condition. The loss in stability leading to a hard flutter mode may occur within a very narrow speed range, and the straightforward tracing of such parameters as logarithmic decrement of the natural vibrations will not give an adequate warning of danger. Means for overcoming this obstacle are not yet known. One interesting suggestion is to incorporate artificial stability in the test airplane, thus raising the flutter speed of the test configuration to an artificially high value, well above the flight range. To interpret the test data, the effects of the artificial stability are then subtracted from the measurements. The technique has been used successfully in the study of mild control-surface modes, but its application to hard flutter modes is still open to examination.

REFERENCES

1. Goland, M., and Luke, Y. L. "An Exact Solution for Two Dimensional Linear Panel Flutter at Supersonic Speeds," *Jour. Aero. Sciences*, 21 (April, 1954): 275–276.

2. Fung, Y. C. "The Static Stability of a Two-Dimensional Curved Panel in a Supersonic Flow, With an Application to Panel Flutter," *Jour. Aero. Sciences*, **21** (August, 1954): 556–565.
3. Miles, J. W. "On the Aerodynamic Instability of Thin Panels," *Jour. Aero. Sciences*, **23** (August, 1956): 771–780.
4. Runyan, H. L., Cunningham, H. J., and Watkins, C. E. "Theoretical Investigation of Several Types of Single Degree of Freedom Flutter," *Jour. Aero. Sciences*, **19** (February, 1952): 101–110.
5. Rainey, A. G. "Some Observations on Stall Flutter and Buffeting," *NACA Res. Memo*, L53E15 (1953).
6. Rainey, A. G. "Preliminary Study of Some Factors Which Affect the Stall-Flutter Characteristics of Thin Wings," *NACA Tech. Note* 3622 (1956).
7. Niedenfuhr, F. W. "On the Possibility of Aeroelastic Reversal of Propeller Blades," *Jour. Aero. Sciences*, **22** (June, 1955): 438–440.
8. Biot, M. A. "The Divergence of Supersonic Wings Including Chordwise Bending," *Jour. Aero. Sciences*, **23** (March, 1956): 237–251.
9. Byrum, B. L., and Grady, E. R. "General Air-Frame Dynamics of a Guided Missile," *Jour. Aero. Sciences*, **22** (August, 1955): 534–540.
10. Rosenberg, R. M., and Stoner, G. "On the Flight Dynamics of Slender Special-Purpose Aircraft," *Jour. Aero. Sciences*, **19** (January, 1952): 29–38.
11. Freidrich, H. R., and Dore, F. J. "The Dynamic Motion of a Missile Descending Through the Atmosphere," *Jour. Aero. Sciences*, **22** (September, 1955): 628–632.
12. Bolz, R. E. "Dynamic Stability of a Missile in Rolling Flight," *Jour. Aero. Sciences*, **19** (June, 1952): 395–403.
13. Stone, R. W., and Bihrle, W. "Studies of Some Effects of Airplane Configuration on the Response to Longitudinal Control in Landing Approaches," *Jour. Aero. Sciences*, **20** (August, 1953): 555–562.
14. Bollay, W. "Aerodynamic Stability and Automatic Control," *Jour. Aero. Sciences*, **18** (September, 1951): 569–623.
15. Rea, J. B. "Analysis of Systems for Automatic Control of Aircraft," *Aero. Engr'g. Review*, **10** (November, 1951): 39–47.
16. Seamans, R. C., Barnes, F. A., Garber, T. B., and Howard, V. W. "Recent Developments in Aircraft Control," *Jour. Aero. Sciences*, **23** (March, 1955): 145–164.
17. Rea, J. B. "Dynamic Analysis of Aeroelastic Aircraft by the Transfer Function—Fourier Method," *Jour. Aero. Sciences*, **18** (June, 1951): 375–397.
18. Berman, A. "Response Matrix Method of Rotor Blade Analysis," *Jour. Aero. Sciences*, **23** (February, 1956): 155–161.
19. Shulman, Y. "Stability of a Flexible Helicopter Rotor Blade in Forward Flight," *Jour. Aero. Sciences*, **23** (July, 1956): 663–670.
20. Coleman, R. P., and Feingold, A. M. "Theory of Self-Excited Mechanical Oscillations of Helicopter Rotors with Hinged Blades," *NACA Tech. Note* 3844 (1957).
21. Kaufman, L., and Peress, K. "A Review of Methods of Predicting Helicopter Longitudinal Response," *Jour. Aero. Sciences*, **23** (March, 1956): 259–271.
22. Milwitzky, B. "Generalized Theory for Seaplane Impact," *NACA Rept.* 1103 (1953).
23. Schnitzer, E. "Theory and Procedure for Determining Loads and Motions in Chine-immersed Hydrodynamic Impacts of Presimatic Bodies," *NACA Rept.* 1152 (1953).
24. Mottard, E. J. "Preliminary Investigation of Self-Excited Vibrations of Single Planing Surfaces," *NACA Tech. Note* 3698 (1956).
25. Kelley, H. J. "Prediction of Yawing Stability Characteristics of Airplanes During Catapulting," *Jour. Aero. Sciences*, **19** (August, 1952): 529–539.
26. Oswald, T. W. "Dynamic Behavior During Accelerated Flight with Particular Application to Missile Launching," *Jour. Aero. Sciences*, **23** (August, 1956): 781–791.
27. Moreland, W. J. "The Story of Shimmy," *Jour. Aero. Sciences*, **21** (December, 1954): 793–808.

28. Regier, A. A. "Noise, Vibration, and Aircraft Structures," *Aero. Engr'g. Review,* **16** (August, 1956): 56–61.
29. Lassiter, L. W., Hess, R. W., and Hubbard, H. H. "An Experimental Study of the Response of Simple Panels to Intense Acoustic Loading," *Jour. Aero. Sciences,* **24** (January, 1957): 19–24.
30. Boley, B. A., and Barber, A. D. "Dynamic Response of Beams and Plates to Rapid Heating," *Jour. App. Mechs.,* **24** (September, 1957): 413–416.
31. Woolston, D. S., Runyan, H. L., and Anderson, R. E. "An Investigation of Effects of Certain Types of Structural Nonlinearities on Wing and Control Surface Flutter," *Jour. Aero. Sciences,* **24** (January, 1957): 57–63.
32. Schade, R. O., Gates, O. B., and Hassell, J. L. "The Effects on Dynamic Lateral Stability and Controllability of Large Artificial Variations in the Rotary Stability Derivatives," *Jour. Aero. Sciences,* **19** (September, 1952): 601–608.
33. Oswald, T. W. "The Effect of Nonlinear Aerodynamics Characteristics on the Dynamic Response to a Sudden Change in Angle of Attack," *Jour. Aero. Sciences,* **19** (May, 1952): 302–316.
34. Covert, E. E. "A Particular Solution of a Nonlinear Differential Equation Which Occurs in the Dynamics of Conservative Systems and Its Application to Undamped Airplane Motion," *Jour. Aero. Sciences,* **22** (February, 1955): 134–135.
35. Curtman, H. J. "Theoretical and Analog Studies of the Effects of Nonlinear Stability Derivatives on the Longitudinal Motions of an Aircraft in Response to Step Control Deflections and to the Influence of Proportional Automatic Control," *NACA Rept.* 1241 (1955).
36. Bisplinghoff, R. L., Ashley, H., and Halfman, R. L. *Aeroelasticity.* Cambridge, Mass.: Addison-Wesley Publishing Co., Inc., 1955.
37. Lewis, R. C., and Wrisley, D. I. "A System for the Excitation of Pure Normal Modes of Complex Structure," *Jour. Aero. Sciences,* **17** (November, 1950): 705–722.
38. Kinnaman, E. B. "Flutter Analysis of Complex Airplanes by Experimental Methods," *Jour. Aero. Sciences,* **19** (September, 1952): 577–584.
39. McCann, G. D., and MacNeal, R. H. "Beam Vibration Analysis with the Electric Analog Computer," *Jour. App. Mechs.,* **17** (March, 1950): 13–26.
40. Benscoter, S. U., and MacNeal, R. H. "Introduction to Electrical-Circuit Analogies for Beam Analysis," *NACA Tech. Note* 2785 (1952).
41. MacNeal, R. H., McCann, G. D., and Wilts, C. H. "The Solution of Aeroelastic Problems by Means of Electrical Analogies," *Jour. Aero. Sciences,* **18** (December, 1951): 777–789.
42. Winston, J. "The Solution of Aeroelastic Problems by Electronic Analog Computations," *Jour. Aero. Sciences,* **17** (July, 1950): 385–395.
43. Scanlan, R. H. "A Steady-Flow Aeroelastic Study by Electrical Analogy," *Jour. Aero. Sciences,* **20** (October, 1953): 691–698.
44. Fung, Y. C. *An Introduction to the Theory of Aeroelasticity.* New York: John Wiley & Sons, Inc., 1955.
45. Serbin, H., and Castilow, E. L. "Application of Response Functions to Calculation of Flutter Characteristics of a Wing Carrying Concentrated Masses," *NACA Tech. Note* 2540 (1951).
46. Goland, M. "The Flutter of a Uniform Cantilever Wing," *Jour. App. Mechs.,* **12** (December, 1945): 197–208.
47. Fraejis de Veubeke, B. M. "A Variational Approach to Pure Mode Excitation Based on Characteristic Lag Theory." A paper prepared for Structures and Materials Panel of AGARD, 1956.
48. Garrick, I. E. "Aerodynamic Theory and Its Application to Flutter." A paper prepared for Structures and Materials Panel of AGARD, 1956.
49. Hedgepeth, J. M. "Recent Research on the Determination of Natural Modes and Frequencies of Aircraft Wing Structures." A paper prepared for Structures and Materials Panel of AGARD, 1956.

APPENDIX A

DERIVATION OF LAGRANGE'S EQUATION

Consider a system of particles in space acted upon by a resultant force whose components are X, Y, Z. The equations of motion are

$$\sum m \frac{d^2x}{dt^2} = X$$

$$\sum m \frac{d^2y}{dt^2} = Y \qquad \text{(A–1)}$$

$$\sum m \frac{d^2z}{dt^2} = Z$$

The *potential energy* of the system, U, may be obtained from the principle that the change in potential energy is equal to the negative of the work done:

$$dU = -(X\,dx + Y\,dy + Z\,dz) \qquad \text{(A–2)}$$

If we call the *generalized coordinates** q_i, then an infinitesimal displacement dx may be written as the total differential:

$$dx = \frac{\partial x}{\partial q_1}\,dq_1 + \frac{\partial x}{\partial q_2}\,dq_2 + \cdots + \frac{\partial x}{\partial q_n}\,dq_n \qquad \text{(A–3)}$$

The change in potential energy then becomes

$$\begin{aligned} -dU = X&\left[\frac{\partial x}{\partial q_1}\,dq_1 + \frac{\partial x}{\partial q_2}\,dq_2 + \cdots + \frac{\partial x}{\partial q_n}\,dq_n\right] \\ &+ Y\left[\frac{\partial y}{\partial q_1}\,dq_1 + \frac{\partial y}{\partial q_2}\,dq_2 + \cdots + \frac{\partial y}{\partial q_n}\,dq_n\right] \\ &+ Z\left[\frac{\partial z}{\partial q_1}\,dq_1 + \frac{\partial z}{\partial q_2}\,dq_2 + \cdots + \frac{\partial z}{\partial q_n}\,dq_n\right] \end{aligned}$$

The potential energy is found, by integrating, as

$$\begin{aligned} -U = &\int\left(X\frac{\partial x}{\partial q_1} + Y\frac{\partial y}{\partial q_1} + Z\frac{\partial z}{\partial q_1}\right)dq_1 \\ &+ \int\left(X\frac{\partial x}{\partial q_2} + Y\frac{\partial y}{\partial q_2} + Z\frac{\partial z}{\partial q_2}\right)dq_2 + \cdots \\ &+ \int\left(X\frac{\partial x}{\partial q_n} + Y\frac{\partial y}{\partial q_n} + Z\frac{\partial z}{\partial q_n}\right)dq_n \end{aligned}$$

* Any system of completely *independent* coordinates is called a *generalized coordinate* system.

We now idealize the problem to the extent that the resultant force acting on the system is derivable from a potential function. That is, the derivative of the potential function in a certain direction gives the component of the force in that direction. Thus, for a given generalized coordinate q_r, we have (from the preceding equation)

$$-\frac{\partial U}{\partial q_r} = X\frac{\partial x}{\partial q_r} + Y\frac{\partial y}{\partial q_r} + Z\frac{\partial z}{\partial q_r} \tag{A-4}$$

Now consider Eq. A-1. Multiplying the first by $\partial x/\partial q_r$, the second by $\partial y/\partial q_r$, and the third by $\partial z/\partial q_r$, adding and combining with Eq. A-4 gives

$$\left(\sum m\frac{d^2x}{dt^2}\right)\frac{\partial x}{\partial q_r} + \left(\sum m\frac{d^2y}{dt^2}\right)\frac{\partial y}{\partial q_r} + \left(\sum m\frac{d^2z}{dt^2}\right)\frac{\partial z}{\partial q_r} = -\frac{\partial U}{\partial q_r} \tag{A-5}$$

The first term on the left-hand side of this equation may be rewritten, if we first note that

$$\frac{d}{dt}\left(\frac{dx}{dt}\frac{\partial x}{\partial q_r}\right) = \frac{\partial x}{\partial q_r}\frac{d}{dt}\left(\frac{dx}{dt}\right) + \frac{dx}{dt}\frac{d}{dt}\left(\frac{\partial x}{\partial q_r}\right)$$

Then

$$\frac{d^2x}{dt^2}\frac{\partial x}{\partial q_r} = \frac{d}{dt}\left(\frac{dx}{dt}\frac{\partial x}{\partial q_r}\right) - \frac{dx}{dt}\frac{d}{dt}\left(\frac{\partial x}{\partial q_r}\right)$$

and similarly

$$\frac{d^2y}{dt^2}\frac{\partial y}{\partial q_r} = \frac{d}{dt}\left(\frac{dy}{dt}\frac{\partial y}{\partial q_r}\right) - \frac{dy}{dt}\frac{d}{dt}\left(\frac{\partial y}{\partial q_r}\right)$$

$$\frac{d^2z}{dt^2}\frac{\partial z}{\partial q_r} = \frac{d}{dt}\left(\frac{dz}{dt}\frac{\partial z}{\partial q_r}\right) - \frac{dz}{dt}\frac{d}{dt}\left(\frac{\partial z}{\partial q_r}\right)$$

Introducing these relations into Eq. A-5 gives

$$\begin{aligned}\frac{d}{dt}\sum m\left(\frac{dx}{dt}\frac{\partial x}{\partial q_r} + \frac{dy}{dt}\frac{\partial y}{\partial q_r} + \frac{dz}{dt}\frac{\partial z}{\partial q_r}\right)\\ - \sum m\left[\frac{dx}{dt}\frac{d}{dt}\left(\frac{\partial x}{\partial q_r}\right) + \frac{dy}{dt}\frac{d}{dt}\left(\frac{\partial y}{\partial q_r}\right) + \frac{dz}{dt}\frac{d}{dt}\left(\frac{\partial z}{\partial q_r}\right)\right] = -\frac{\partial U}{\partial q_r}\end{aligned} \tag{A-6}$$

The kinetic energy of the system is

$$T = \tfrac{1}{2}\sum m\left[\left(\frac{dx}{dt}\right)^2 + \left(\frac{dy}{dt}\right)^2 + \left(\frac{dz}{dt}\right)^2\right] \tag{A-7}$$

from which

$$\frac{\partial T}{\partial q_r} = \sum m\left[\frac{dx}{dt}\frac{\partial}{\partial q_r}\left(\frac{dx}{dt}\right) + \frac{dy}{dt}\frac{\partial}{\partial q_r}\left(\frac{dy}{dt}\right) + \frac{dz}{dt}\frac{\partial}{\partial q_r}\left(\frac{dz}{dt}\right)\right] \tag{A-8}$$

$$\frac{\partial T}{\partial \dot{q}_r} = \sum m\left[\frac{dx}{dt}\frac{\partial}{\partial \dot{q}_r}\left(\frac{dx}{dt}\right) + \frac{dy}{dt}\frac{\partial}{\partial \dot{q}_r}\left(\frac{dy}{dt}\right) + \frac{dz}{dt}\frac{\partial}{\partial \dot{q}_r}\left(\frac{dz}{dt}\right)\right] \tag{A-9}$$

Now, from Eq. A–3 we may write

$$\dot{x} = \frac{dx}{dt} = \frac{\partial x}{\partial q_1}\dot{q}_1 + \frac{\partial x}{\partial q_2}\dot{q}_2 + \cdots + \frac{\partial x}{\partial q_n}\dot{q}_n \tag{A–10}$$

so that

$$\frac{\partial \dot{x}}{\partial \dot{q}_r} = \frac{\partial x}{\partial q_r} \tag{A–11}$$

Also, from Eqs. A–3 and A–10 we may write

$$\frac{\partial \dot{x}}{\partial q_r} = \frac{\partial^2 x}{\partial q_1 \partial q_r}\dot{q}_1 + \frac{\partial^2 x}{\partial q_2 \partial q_r}\dot{q}_2 + \cdots + \frac{\partial^2 x}{\partial q_n \partial q_r}\dot{q}_n$$

$$\frac{d}{dt}\left(\frac{\partial x}{\partial q_r}\right) = \frac{\partial^2 x}{\partial q_1 \partial q_r}\dot{q}_1 + \frac{\partial^2 x}{\partial q_2 \partial q_r}\dot{q}_2 + \cdots + \frac{\partial^2 x}{\partial q_n \partial q_r}\dot{q}_n$$

Therefore

$$\frac{\partial \dot{x}}{\partial q_r} = \frac{d}{dt}\left(\frac{\partial x}{\partial q_r}\right) \tag{A–12}$$

Results similar to Eqs. A–11 and A–12 may be obtained for the variables y and z.

Now introduce Eq. A–12 and the similar expressions for y and z into Eq. A–8, and compare this result with the negative term on the left-hand side of Eq. A–6. Therefore, Eq. A–6 becomes

$$\frac{d}{dt}\sum m\left(\frac{dx}{dt}\frac{\partial x}{\partial q_r} + \frac{dy}{dt}\frac{\partial y}{\partial q_r} + \frac{dz}{dt}\frac{\partial z}{\partial q_r}\right) - \frac{\partial T}{\partial q_r} = -\frac{\partial U}{\partial q_r} \tag{A–13}$$

Now introduce Eq. A–11 into Eq. A–9, and compare with the first term on the left-hand side of Eq. A–13. Therefore, Eq. A–13 becomes

$$\frac{d}{dt}\left(\frac{\partial T}{\partial \dot{q}_r}\right) - \frac{\partial T}{\partial q_r} + \frac{\partial U}{\partial q_r} = 0$$

This result is *Lagrange's equation* as given in the main text (see Eq. 2–8).

APPENDIX B

ORTHOGONAL MODES OF VIBRATION

We have seen that the general expression for the vibration of a simply supported beam may be expressed by a series of sine terms the coefficients of which are functions of time (Eq. 2–54). This is a special case of the expansion of a function in a series of *orthogonal* functions.

Two functions $\alpha_1(x)$ and $\alpha_2(x)$ are said to be *orthogonal* if in the region $x = a$ to $x = b$,

$$\int_a^b \alpha_1 \, \alpha_2 \, dx = 0 \tag{B–1}$$

A function $\alpha_n(x)$ is said to be *normalized* if

$$\int_a^b \alpha_n{}^2 \, dx = 1 \tag{B–2}$$

Therefore, a set of normalized orthogonal functions has the property that

$$\begin{aligned} \int_a^b \alpha_r \alpha_s \, dx = \delta_{rs} &= 1, \quad (r = s) \\ &= 0, \quad (r \neq s) \end{aligned} \tag{B–3}$$

The coefficients of a given series of orthogonal functions

$$S(x) = A_1\alpha_1(x) + A_2\alpha_2(x) + \cdots + A_n\alpha_n(x) + \cdots \tag{B–4}$$

may be evaluated by means of the orthogonality and normalization conditions. If we multiply each side of the above series by $\alpha_n(x)$ and introduce the relation of Eq. B–3, we obtain

$$A_n = \int_a^b S(x)\alpha_n(x) \, dx \tag{B–5}$$

It is possible to interpret the orthogonality condition geometrically in the case of the principal vibrations of an elastic system.* In particular it can be shown that the vibration of a particle in space can always be made up by superposition of vibrations in three mutually perpendicular directions. If we were to set up such a problem in such a manner that the coordinate axes were in the directions of the three principal vibrations, then each equation of motion would contain only one

* For a complete discussion, see T. von Kármán and M. A. Biot, *Mathematical Methods in Engineering.* (New York: McGraw-Hill Book Co., Inc.), 1940; pp. 172–178.

coordinate, and we would therefore have the vibration properties expressed directly in terms of the normal modes.

The use of the orthogonality condition is often of great help if one desires to represent some arbitrary vibration shape by a superposition of normal modes. For example, in the case of the simply supported beam undergoing free vibrations, we have found that the deflection could be described as the sum of the normal modes of vibration as (see Eq. 2–54)

$$y = \sum_{n=1}^{\infty} (A_n \cos \omega_n t + B_n \sin \omega_n t) \sin \frac{n\pi x}{l} \tag{B-6}$$

If now it were desired to represent the arbitrary deflection curve given by y_1, y_2, $\cdots$ y_n (the subscripts refer to stations along the span of the beam), then there would be the n equations

$$\begin{aligned} y_1 &= c_1\xi_1^{(1)} + c_2\xi_1^{(2)} + \cdots \\ y_2 &= c_1\xi_2^{(1)} + c_2\xi_2^{(2)} + \cdots \\ &\cdot \\ &\cdot \\ y_n &= c_1\xi_n^{(1)} + c_2\xi_n^{(2)} + \cdots \end{aligned} \tag{B-7}$$

where, for example, $\xi_1^{(1)}$ represents the amplitude of the first normal mode (superscript) at the first station (subscript). The constants c_1, c_2, $\cdots$ c_n, which define the magnitudes of the contributions of the various modes, may be found by application of the orthogonality and normalization conditions (Eq. B–3) in a manner similar to that used in the series of Eq. B–4 to obtain the coefficients in the form of Eq. B–5.

The term orthogonality comes originally from vector analysis, where it is shown that the scalar (dot) product of two vectors is zero if the vectors are at right angles to each other.

APPENDIX C

SOLUTIONS TO PROBLEMS

Note: Since a substantial number of problems in the text involve the derivation of proofs or contain the answer to the problem statement, this answer list actually covers about two-thirds of all the problems.

CHAPTER 1

1–1. $c = 0.60 \dfrac{\#\,\text{sec}}{\text{ft}}$

1–3. $t_{max} = \dfrac{1}{2\sqrt{n^2 \omega^2}} \ln \left[\dfrac{-n - \sqrt{n^2 - \omega^2}}{-n + \sqrt{n^2 - \omega^2}}\right]$

1–5. $\tau = 0.72$ sec $\qquad x_{max} = 5$ in.

1–6. $f = \dfrac{1}{2\pi} \sqrt{\dfrac{k_1 k_2}{m(k_1 + k_2)}}$

1–7. $f = \dfrac{1}{2\pi} \sqrt{\dfrac{2k}{m}}$

1–8. $W_2 = \left(\dfrac{\tau_2}{\tau_1}\right)^2 W_1$

1–9. $x = x_0 \cos \omega t + \left(\dfrac{v_0}{\omega} - \dfrac{K}{m\omega^3}\right) \sin \omega t + \dfrac{Kt}{m\omega^2}$

1–10. $\ddot{\theta} + \dfrac{g}{l} \sin \theta = 0$; for small θ, $\qquad \ddot{\theta} + \dfrac{g}{l} \theta = 0,$

$$\theta = \theta_0 \cos \sqrt{\frac{g}{l}}\, t + \frac{\dot{\theta}_0}{\omega} \sin \sqrt{\frac{g}{l}}\, t$$

1–11. (a) $f = 1.72$ cps (b) $f = 1.62$ cps

1–12. $\theta = 2.6^\circ$

1–13. $|MF| = (1 - \eta^4)^{-1/2}$

1–14. $m\ddot{x} + c\dot{x} + kx = kX_0 \cos \Omega t$

CHAPTER 2

2–2. $f = \dfrac{1}{2\pi}\left[\dfrac{GJ}{lI_1 I_2}(I_1 + I_2)\right]^{1/2}$

2–3. $I_1 \ddot{\theta}_1 + \dfrac{GJ}{l}(\theta_1 - \theta_2) = 0 \qquad I_2 \ddot{\theta}_2 + \dfrac{GJ}{l}(\theta_2 - \theta_1) = 0$

2–4. $\omega_1 = \sqrt{g/l} \qquad \omega_2 = \sqrt{\dfrac{2kh^2}{ml^2} + g/l}$

2–5. $(m_2 + m_1) l_1^2 \ddot{\varphi}_1 + m_2 l_1 l_2 \ddot{\varphi}_2 + (m_1 g + m_2 g) l_1 \varphi_1 = 0$

$m_2 (l_2^2 \ddot{\varphi}_2 + l_1 l_2 \ddot{\varphi}_1) + m_2 g l_2 \varphi_2 = 0$

2–6. $I_1 \ddot{\varphi}_1 + \left(\dfrac{GJ}{l_1} + \dfrac{GJ}{l_2}\right)\varphi_1 - \dfrac{GJ}{l_2}\varphi_2 = 0$

$I_2 \ddot{\varphi}_2 + \dfrac{GJ}{l_2}\varphi_2 - \dfrac{GJ}{l_2}\varphi_1 = 0$

2–7. $m_1 \ddot{x}_1 + c(\dot{x}_1 - \dot{x}_2) + k_1(x_1 - x_2) = 0$

$m_2 \ddot{x}_2 - c(\dot{x}_1 - \dot{x}_2) - k_1(x_1 - x_2) + k_2 x_2 = 0$

2–8. $M\ddot{h} + S\ddot{\alpha} + k_1 h = 0$ $\qquad M$ = wing mass

$S\ddot{h} + I\ddot{\alpha} + k_2 \alpha = 0$ $\qquad I$ = mass moment of inertia

S = static moment

2–11. $14.1 \sqrt{\dfrac{EI}{l^4 w/g}}$ cyc/min

2–13. $w_i = \dfrac{a_i}{l^2}\sqrt{\dfrac{EI}{m}}, \qquad a_1 = 3.515, \qquad a_2 = 22.03, \qquad a_3 = 61.7$

2–14. $\sin kl = 0$

2–15. $\cos kl \cosh kl = 1$

CHAPTER 3

3–1. -402

3–2. $[a] + [b] = \begin{bmatrix} 3 & 5 \\ 8 & -4 \end{bmatrix}$ $\quad [a][b] = \begin{bmatrix} 12 & 3 \\ -14 & 9 \end{bmatrix}$

$[a] - [b] = \begin{bmatrix} -1 & -1 \\ -2 & -4 \end{bmatrix}$ $\quad [b][a] = \begin{bmatrix} 11 & -8 \\ 5 & 10 \end{bmatrix}$

3–3. $[c]' = \begin{bmatrix} -2 & 1 & 3 \\ 3 & 0 & -1 \\ 4 & 2 & 5 \end{bmatrix}$ $\quad [c]^{-1} = -\frac{1}{5}\begin{bmatrix} 2 & -19 & 6 \\ 1 & -22 & 8 \\ -1 & 7 & -3 \end{bmatrix}$

3–4. $[M] = \begin{bmatrix} m & 0 \\ 0 & m \end{bmatrix}$ $\quad [K] = \begin{bmatrix} 2k & -k \\ -k & 2k \end{bmatrix}$

$[\Phi] = \frac{1}{3k}\begin{bmatrix} 2 & 1 \\ 1 & 2 \end{bmatrix}$

CHAPTER 4

4–2. $4k < \frac{c^2}{m - P}$ unstable

4–3. $\alpha = 14.1^\circ$

CHAPTER 7

7–2. $(D_n)_{max} = 1 + \frac{1}{\omega_n t_1}\sqrt{2(1 - \cos \omega_n t_1)}$

7–3. $u(x, t) = \frac{gF_0}{A\,lw/g}\sum_{n=1}^{\infty}\frac{K_n}{\omega_n^2}\varphi_n(x)D_n$

INDEX

A CATALOGUE OF SELECTED DOVER SCIENCE BOOKS

A CATALOGUE OF SELECTED DOVER SCIENCE BOOKS

Physics: The Pioneer Science, Lloyd W. Taylor. Very thorough non-mathematical survey of physics in a historical framework which shows development of ideas. Easily followed by laymen; used in dozens of schools and colleges for survey courses. Richly illustrated. Volume 1: Heat, sound, mechanics. Volume 2: Light, electricity. Total of 763 illustrations. Total of cvi + 847pp.
60565-5, 60566-3 Two volumes, Paperbound 5.50

THE RISE OF THE NEW PHYSICS, A. d'Abro. Most thorough explanation in print of central core of mathematical physics, both classical and modern, from Newton to Dirac and Heisenberg. Both history and exposition: philosophy of science, causality, explanations of higher mathematics, analytical mechanics, electromagnetism, thermodynamics, phase rule, special and general relativity, matrices. No higher mathematics needed to follow exposition, though treatment is elementary to intermediate in level. Recommended to serious student who wishes verbal understanding. 97 illustrations. Total of ix + 982pp.
20003-5, 20004-3 Two volumes, Paperbound $5.50

INTRODUCTION TO CHEMICAL PHYSICS, John C. Slater. A work intended to bridge the gap between chemistry and physics. Text divided into three parts: Thermodynamics, Statistical Mechanics, and Kinetic Theory; Gases, Liquids and Solids; and Atoms, Molecules and the Structure of Matter, which form the basis of the approach. Level is advanced undergraduate to graduate, but theoretical physics held to minimum. 40 tables, 118 figures. xiv + 522pp.
62562-1 Paperbound $4.00

BASIC THEORIES OF PHYSICS, Peter C. Bergmann. Critical examination of important topics in classical and modern physics. Exceptionally useful in examining conceptual framework and methodology used in construction of theory. Excellent supplement to any course, textbook. Relatively advanced.
Volume 1. Heat and Quanta. Kinetic hypothesis, physics and statistics, stationary ensembles, thermodynamics, early quantum theories, atomic spectra, probability waves, quantization in wave mechanics, approximation methods, abstract quantum theory. 8 figures. x + 300pp. 60968-5 Paperbound $2.00
Volume 2. Mechanics and Electrodynamics. Classical mechanics, electro- and magnetostatics, electromagnetic induction, field waves, special relativity, waves, etc. 16 figures, viii + 260pp. 60969-3 Paperbound $2.75

FOUNDATIONS OF PHYSICS, Robert Bruce Lindsay and Henry Margenau. Methods and concepts at the heart of physics (space and time, mechanics, probability, statistics, relativity, quantum theory) explained in a text that bridges gap between semi-popular and rigorous introductions. Elementary calculus assumed. "Thorough and yet not over-detailed," *Nature*. 35 figures. xviii + 537 pp.
60377-6 Paperbound $3.50

FUNDAMENTAL FORMULAS OF PHYSICS, edited by Donald H. Menzel. Most useful reference and study work, ranges from simplest to most highly sophisticated operations. Individual chapters, with full texts explaining formulae, prepared by leading authorities cover basic mathematical formulas, statistics, nomograms, physical constants, classical mechanics, special theory of relativity, general theory of relativity, hydrodynamics and aerodynamics, boundary value problems in mathematical physics, heat and thermodynamics, statistical mechanics, kinetic theory of gases, viscosity, thermal conduction, electromagnetism, electronics, acoustics, geometrical optics, physical optics, electron optics, molecular spectra, atomic spectra, quantum mechanics, nuclear theory, cosmic rays and high energy phenomena, particle accelerators, solid state, magnetism, etc. Special chapters also cover physical chemistry, astrophysics, celestian mechanics, meteorology, and biophysics. Indispensable part of library of every scientist. Total of xli + 787pp.
60595-7, 60596-5 Two volumes, Paperbound $5.00

INTRODUCTION TO EXPERIMENTAL PHYSICS, William B. Fretter. Detailed coverage of techniques and equipment: measurements, vacuum tubes, pulse circuits, rectifiers, oscillators, magnet design, particle counters, nuclear emulsions, cloud chambers, accelerators, spectroscopy, magnetic resonance, x-ray diffraction, low temperature, etc. One of few books to cover laboratory hazards, design of exploratory experiments, measurements. 298 figures. xii + 349pp.
(EUK) 61890-0 Paperbound $2.50

CONCEPTS AND METHODS OF THEORETICAL PHYSICS, Robert Bruce Lindsay. Introduction to methods of theoretical physics, emphasizing development of physical concepts and analysis of methods. Part I proceeds from single particle to collections of particles to statistical method. Part II covers application of field concept to material and non-material media. Numerous exercises and examples. 76 illustrations. x + 515pp. 62354-8 Paperbound $4.00

AN ELEMENTARY TREATISE ON THEORETICAL MECHANICS, Sir James Jeans. Great scientific expositor in remarkably clear presentation of basic classical material: rest, motion, forces acting on particle, statics, motion of particle under variable force, motion of rigid bodies, coordinates, etc. Emphasizes explanation of fundamental physical principles rather than mathematics or applications. Hundreds of problems worked in text. 156 figures. x + 364pp. 61839-0 Paperbound $2.50

THEORETICAL MECHANICS: AN INTRODUCTION TO MATHEMATICAL PHYSICS, Joseph S. Ames and Francis D. Murnaghan. Mathematically rigorous introduction to vector and tensor methods, dynamics, harmonic vibrations, gyroscopic theory, principle of least constraint, Lorentz-Einstein transformation. 159 problems; many fully-worked examples. 39 figures. ix + 462pp. 60461-6 Paperbound $3.00

THE PRINCIPLE OF RELATIVITY, Albert Einstein, Hendrick A. Lorentz, Hermann Minkowski and Hermann Weyl. Eleven original papers on the special and general theory of relativity, all unabridged. Seven papers by Einstein, two by Lorentz, one each by Minkowski and Weyl. "A thrill to read again the original papers by these giants," *School Science and Mathematics.* Translated by W. Perret and G. B. Jeffery. Notes by A. Sommerfeld. 7 diagrams. viii + 216pp.
60081-5 Paperbound $2.00

EINSTEIN'S THEORY OF RELATIVITY, Max Born. Relativity theory analyzed, explained for intelligent layman or student with some physical, mathematical background. Includes Lorentz, Minkowski, and others. Excellent verbal account for teachers. Generally considered the finest non-technical account. vii + 376pp.
60769-0 Paperbound $2.50

PHYSICAL PRINCIPLES OF THE QUANTUM THEORY, Werner Heisenberg. Nobel Laureate discusses quantum theory, uncertainty principle, wave mechanics, work of Dirac, Schroedinger, Compton, Wilson, Einstein, etc. Middle, non-mathematical level for physicist, chemist not specializing in quantum; mathematical appendix for specialists. Translated by C. Eckart and F. Hoyt. 19 figures. viii + 184pp. 60113-7 Paperbound $2.00

PRINCIPLES OF QUANTUM MECHANICS, William V. Houston. For student with working knowledge of elementary mathematical physics; uses Schroedinger's wave mechanics. Evidence for quantum theory, postulates of quantum mechanics, applications in spectroscopy, collision problems, electrons, similar topics. 21 figures. 288pp. 60524-8 Paperbound $3.00

ATOMIC SPECTRA AND ATOMIC STRUCTURE, Gerhard Herzberg. One of the best introductions to atomic spectra and their relationship to structure; especially suited to specialists in other fields who require a comprehensive basic knowledge. Treatment is physical rather than mathematical. 2nd edition. Translated by J. W. T. Spinks. 80 illustrations. xiv + 257pp. 60115-3 Paperbound $2.00

ATOMIC PHYSICS: AN ATOMIC DESCRIPTION OF PHYSICAL PHENOMENA, Gaylord P. Harnwell and William E. Stephens. One of the best introductions to modern quantum ideas. Emphasis on the extension of classical physics into the realms of atomic phenomena and the evolution of quantum concepts. 156 problems. 173 figures and tables. xi + 401pp. 61584-7 Paperbound $2.50

ATOMS, MOLECULES AND QUANTA, Arthur E. Ruark and Harold C. Urey. 1964 edition of work that has been a favorite of students and teachers for 30 years. Origins and major experimental data of quantum theory, development of concepts of atomic and molecular structure prior to new mechanics, laws and basic ideas of quantum mechanics, wave mechanics, matrix mechanics, general theory of quantum dynamics. Very thorough, lucid presentation for advanced students. 230 figures. Total of xxiii + 810pp.
61106-X, 61107-8 Two volumes, Paperbound $6.00

INVESTIGATIONS ON THE THEORY OF THE BROWNIAN MOVEMENT, Albert Einstein. Five papers (1905-1908) investigating the dynamics of Brownian motion and evolving an elementary theory of interest to mathematicians, chemists and physical scientists. Notes by R. Fürth, the editor, discuss the history of study of Brownian movement, elucidate the text and analyze the significance of the papers. Translated by A. D. Cowper. 3 figures. iv + 122pp.
60304-0 Paperbound $1.50

MATHEMATICAL FOUNDATIONS OF STATISTICAL MECHANICS, A. I. Khinchin. Introduction to modern statistical mechanics: phase space, ergodic problems, theory of probability, central limit theorem, ideal monatomic gas, foundation of thermodynamics, dispersion and distribution of sum functions. Provides mathematically rigorous treatment and excellent analytical tools. Translated by George Gamow. viii + 179pp. 60147-1 Paperbound $2.00

INTRODUCTION TO PHYSICAL STATISTICS, Robert B. Lindsay. Elementary probability theory, laws of thermodynamics, classical Maxwell-Boltzmann statistics, classical statistical mechanics, quantum mechanics, other areas of physics that can be studied statistically. Full coverage of methods; basic background theory. ix + 306pp. 61882-X Paperbound $2.75

DIALOGUES CONCERNING TWO NEW SCIENCES, Galileo Galilei. Written near the end of Galileo's life and encompassing 30 years of experiment and thought, these dialogues deal with geometric demonstrations of fracture of solid bodies, cohesion, leverage, speed of light and sound, pendulums, falling bodies, accelerated motion, etc. Translated by Henry Crew and Alfonso de Salvio. Introduction by Antonio Favaro. xxiii + 300pp. 60099-8 Paperbound $2.25

FOUNDATIONS OF SCIENCE: THE PHILOSOPHY OF THEORY AND EXPERIMENT, Norman R. Campbell. Fundamental concepts of science examined on middle level: acceptance of propositions and axioms, presuppositions of scientific thought, scientific law, multiplication of probabilities, nature of experiment, application of mathematics, measurement, numerical laws and theories, error, etc. Stress on physics, but holds for other sciences. "Unreservedly recommended," *Nature* (England). Formerly *Physics: The Elements.* ix + 565pp. 60372-5 Paperbound $4.00

THE PHASE RULE AND ITS APPLICATIONS, Alexander Findlay, A. N. Campbell and N. O. Smith. Findlay's well-known classic, updated (1951). Full standard text and thorough reference, particularly useful for graduate students. Covers chemical phenomena of one, two, three, four and multiple component systems. "Should rank as the standard work in English on the subject," *Nature.* 236 figures. xii + 494pp. 60091-2 Paperbound $3.50

THERMODYNAMICS, Enrico Fermi. A classic of modern science. Clear, organized treatment of systems, first and second laws, entropy, thermodynamic potentials, gaseous reactions, dilute solutions, entropy constant. No math beyond calculus is needed, but readers are assumed to be familiar with fundamentals of thermometry, calorimetry. 22 illustrations. 25 problems. x + 160pp. 60361-X Paperbound $2.00

TREATISE ON THERMODYNAMICS, Max Planck. Classic, still recognized as one of the best introductions to thermodynamics. Based on Planck's original papers, it presents a concise and logical view of the entire field, building physical and chemical laws from basic empirical facts. Planck considers fundamental definitions, first and second principles of thermodynamics, and applications to special states of equilibrium. Numerous worked examples. Translated by Alexander Ogg. 5 figures. xiv + 297pp. 60219-2 Paperbound $2.50

MICROSCOPY FOR CHEMISTS, Harold F. Schaeffer. Thorough text; operation of microscope, optics, photomicrographs, hot stage, polarized light, chemical procedures for organic and inorganic reactions. 32 specific experiments cover specific analyses: industrial, metals, other important subjects. 136 figures. 264pp.
61682-7 Paperbound $2.50

THE ELECTRONIC THEORY OF ACIDS AND BASES, by William F. Luder and Saverio Zuffanti. Full, newly revised (1961) presentation of a still controversial theory. Historical background, atomic orbitals and valence, electrophilic and electrodotic reagents, acidic and basic radicals, titrations, displacement, acid catalysis, etc., are discussed. xi + 165pp. 60201-X Paperbound $2.00

OPTICKS, Sir Isaac Newton. A survey of 18th-century knowledge on all aspects of light as well as a description of Newton's experiments with spectroscopy, colors, lenses, reflection, refraction, theory of waves, etc. in language the layman can follow. Foreword by Albert Einstein. Introduction by Sir Edmund Whittaker. Preface by I. Bernard Cohen. cxxvi + 406pp. 60205-2 Paperbound $3.50

LIGHT: PRINCIPLES AND EXPERIMENTS, George S. Monk. Thorough coverage, for student with background in physics and math, of physical and geometric optics. Also includes 23 experiments on optical systems, instruments, etc. "Probably the best intermediate text on optics in the English language," *Physics Forum.* 275 figures. xi + 489pp. 60341-5 Paperbound $3.50

PIEZOELECTRICITY: AN INTRODUCTION TO THE THEORY AND APPLICATIONS OF ELECTROMECHANICAL PHENOMENA IN CRYSTALS, Walter G. Cady. Revised 1963 edition of most complete, most systematic coverage of field. Fundamental theory of crystal electricity, concepts of piezoelectricity, including comparisons of various current theories; resonators; oscillators; properties, etc., of Rochelle salt; ferroelectric crystals; applications; pyroelectricity, similar topics. "A great work," *Nature.* Many illustrations. Total of xxx + 840pp.
61094-2, 61095-0 Two volumes, Paperbound $6.00

PHYSICAL OPTICS, Robert W. Wood. A classic in the field, this is a valuable source for students of physical optics and excellent background material for a study of electromagnetic theory. Partial contents: nature and rectilinear propagation of light, reflection from plane and curved surfaces, refraction, absorption and dispersion, origin of spectra, interference, diffraction, polarization, Raman effect, optical properties of metals, resonance radiation and fluorescence of atoms, magneto-optics, electro-optics, thermal radiation. 462 diagrams, 17 plates. xvi + 846pp.
61808-0 Paperbound $4.25

MIRRORS, PRISMS AND LENSES: A TEXTBOOK OF GEOMETRICAL OPTICS, James P. C. Southall. Introductory-level account of modern optical instrument theory, covering unusually wide range: lights and shadows, reflection of light and plane mirrors, refraction, astigmatic lenses, compound systems, aperture and field of optical system, the eye, dispersion and achromatism, rays of finite slope, the microscope, much more. Strong emphasis on earlier, elementary portions of field, utilizing simplest mathematics wherever possible. Problems. 329 figures. xxiv + 806pp. 61234-1 Paperbound $3.75

A Treatise on the Differential Geometry of Curves and Surfaces, Luther P. Eisenhart. Detailed, concrete introductory treatise on differential geometry, developed from author's graduate courses at Princeton University. Thorough explanation of the geometry of curves and surfaces, concentrating on problems most helpful to students. 683 problems, 30 diagrams. xiv + 474pp.
60667-8 Paperbound $2.75

An Essay on the Foundations of Geometry, Bertrand Russell. A mathematical and physical analysis of the place of the a priori in geometric knowledge. Includes critical review of 19th-century work in non-Euclidean geometry as well as illuminating insights of one of the great minds of our time. New foreword by Morris Kline. xx + 201pp. 60233-8 Paperbound $2.00

Introduction to the Theory of Numbers, Leonard E. Dickson. Thorough, comprehensive approach with adequate coverage of classical literature, yet simple enough for beginners. Divisibility, congruences, quadratic residues, binary quadratic forms, primes, least residues, Fermat's theorem, Gauss's lemma, and other important topics. 249 problems, 1 figure. viii + 183pp.
60342-3 Paperbound $2.00

An Elementary Introduction to the Theory of Probability, B. V. Gnedenko and A. Ya. Khinchin. Introduction to facts and principles of probability theory. Extremely thorough within its range. Mathematics employed held to elementary level. Excellent, highly accurate layman's introduction. Translated from the fifth Russian edition by Leo Y. Boron. xii + 130pp.
60155-2 Paperbound $1.75

Selected Papers on Noise and Stochastic Processes, edited by Nelson Wax. Six papers which serve as an introduction to advanced noise theory and fluctuation phenomena, or as a reference tool for electrical engineers whose work involves noise characteristics, Brownian motion, statistical mechanics. Papers are by Chandrasekhar, Doob, Kac, Ming, Ornstein, Rice, and Uhlenbeck. Exact facsimile of the papers as they appeared in scientific journals. 19 figures. v + 337pp. 6⅛ x 9¼.
60262-1 Paperbound $3.00

Statistics Manual, Edwin L. Crow, Frances A. Davis and Margaret W. Maxfield. Comprehensive, practical collection of classical and modern methods of making statistical inferences, prepared by U. S. Naval Ordnance Test Station. Formulae, explanations, methods of application are given, with stress on use. Basic knowledge of statistics is assumed. 21 tables, 11 charts, 95 illustrations. xvii + 288pp.
60599-X Paperbound $2.00

Mathematical Foundations of Information Theory, A. I. Khinchin. Comprehensive introduction to work of Shannon, McMillan, Feinstein and Khinchin, placing these investigations on a rigorous mathematical basis. Covers entropy concept in probability theory, uniqueness theorem, Shannon's inequality, ergodic sources, the E property, martingale concept, noise, Feinstein's fundamental lemma, Shanon's first and second theorems. Translated by R. A. Silverman and M. D. Friedman. iii + 120pp. 60434-9 Paperbound $1.75

A COURSE IN MATHEMATICAL ANALYSIS, Edouard Goursat. *The entire "Cours d'analyse" for students with one year of calculus, offering an exceptionally wide range of subject matter on analysis and applied mathematics. Available for the first time in English. Definitive treatment.*

VOLUME I: Applications to geometry, expansion in series, definite integrals, derivatives and differentials. Translated by Earle R. Hedrick. 52 figures. viii + 548pp. 60554-X Paperbound $3.00

VOLUME II, PART I: Functions of a complex variable, conformal representations, doubly periodic functions, natural boundaries, etc. Translated by Earle R. Hedrick and Otto Dunkel. 38 figures. x + 259pp. 60555-8 Paperbound $2.25

VOLUME II, PART II: Differential equations, Cauchy-Lipschitz method, non-linear differential equations, simultaneous equations, etc. Translated by Earle R. Hedrick and Otto Dunkel. 1 figure. viii + 300pp. 60556-6 Paperbound $2.50

VOLUME III, PART I: Variation of solutions, partial differential equations of the second order. Poincaré's theorem, periodic solutions, asymptotic series, wave propagation, Dirichlet's problem in space, Newtonian potential, etc. Translated by Howard G. Bergmann. 15 figures. x + 329pp. 61176-0 Paperbound $3.00

VOLUME III, PART II: Integral equations and calculus of variations: Fredholm's equation, Hilbert-Schmidt theorem, symmetric kernels, Euler's equation, transversals, extreme fields, Weierstrass's theory, etc. Translated by Howard G. Bergmann. Note on Conformal Representation by Paul Montel. 13 figures. xi + 389pp. 61177-9 Paperbound $3.00

ELEMENTARY STATISTICS: WITH APPLICATIONS IN MEDICINE AND THE BIOLOGICAL SCIENCES, Frederick E. Croxton. Presentation of all fundamental techniques and methods of elementary statistics assuming average knowledge of mathematics only. Useful to readers in all fields, but many examples drawn from characteristic data in medicine and biological sciences. vii + 376pp. 60506-X Paperbound $2.25

ELEMENTS OF THE THEORY OF FUNCTIONS. A general background text that explores complex numbers, linear functions, sets and sequences, conformal mapping. Detailed proofs. Translated by Frederick Bagemihl. 140pp. 60154-4 Paperbound $1.50

THEORY OF FUNCTIONS, PART I. Provides full demonstrations, rigorously set forth, of the general foundations of the theory: integral theorems, series, the expansion of analytic functions. Translated by Federick Bagemihl. vii + 146pp. 60156-0 Paperbound $1.50

INTRODUCTION TO THE THEORY OF FOURIER'S SERIES AND INTEGRALS, Horatio S. Carslaw. A basic introduction to the theory of infinite series and integrals, with special reference to Fourier's series and integrals. Based on the classic Riemann integral and dealing with only ordinary functions, this is an important class text. 84 examples. xiii + 368pp. 60048-3 Paperbound $3.00

AN INTRODUCTION TO FOURIER METHODS AND THE LAPLACE TRANSFORMATION, Philip Franklin. Introductory study of theory and applications of Fourier series and Laplace transforms, for engineers, physicists, applied mathematicians, physical science teachers and students. Only a previous knowledge of elementary calculus is assumed. Methods are related to physical problems in heat flow, vibrations, eletcrical transmission, electromagnetic radiation, etc. 828 problems with answers. Formerly *Fourier Methods.* x + 289pp. 60452-7 Paperbound $2.50

INFINITE SEQUENCES AND SERIES, Konrad Knopp. Careful presentation of fundamentals of the theory by one of the finest modern expositors of higher mathematics. Covers functions of real and complex variables, arbitrary and null sequences, convergence and divergence. Cauchy's limit theorem, tests for infinite series, power series, numerical and closed evaluation of series. Translated by Frederick Bagemihl. v + 186pp. 60153-6 Paperbound $2.00

INTRODUCTION TO THE DIFFERENTIAL EQUATIONS OF PHYSICS, Ludwig Hopf. No math background beyond elementary calculus is needed to follow this classroom or self-study introduction to ordinary and partial differential equations. Approach is through classical physics. Translated by Walter Nef. 48 figures. v + 154pp. 60120-X Paperbound $1.45

DIFFERENTIAL EQUATIONS FOR ENGINEERS, Philip Franklin. For engineers, physicists, applied mathematicians. Theory and application: solution of ordinary differential equations and partial derivatives, analytic functions. Fourier series, Abel's theorem, Cauchy Riemann differential equations, etc. Over 400 problems deal with electricity, vibratory systems, heat, radio; solutions. Formerly *Differential Equations for Electrical Engineers.* 41 illustrations. vii + 299pp. 60601-5 Paperbound $2.00

THEORY OF FUNCTIONS, PART II. Single- and multiple-valued functions; full presentation of the most characteristic and important types. Proofs fully worked out. Translated by Frederick Bagemihl. x + 150pp. 60157-9 Paperbound $1.50

PROBLEM BOOK IN THE THEORY OF FUNCTIONS, I. More than 300 elementary problems for independent use or for use with "Theory of Functions, I." 85pp. of detailed solutions. Translated by Lipman Bers. viii + 126pp. 60158-7 Paperbound $1.50

PROBLEM BOOK IN THE THEORY OF FUNCTIONS, II. More than 230 problems in the advanced theory. Designed to be used with "Theory of Functions, II" or with any comparable text. Full solutions. Translated by Frederick Bagemihl. 138pp. 60159-5 Paperbound $1.50

INTRODUCTION TO THE THEORY OF EQUATIONS, Florian Cajori. Classic introduction by leading historian of science covers the fundamental theories as reached by Gauss, Abel, Galois and Kronecker. Basics of equation study are followed by symmetric functions of roots, elimination, homographic and Tschirnhausen transformations, resolvents of Lagrange, cyclic equations, Abelian equations, the work of Galois, the algebraic solution of general equations, and much more. Numerous exercises include answers. ix + 239pp. 62184-7 Paperbound $2.75

LAPLACE TRANSFORMS AND THEIR APPLICATIONS TO DIFFERENTIAL EQUATIONS, N. W. McLachlan. Introduction to modern operational calculus, applying it to ordinary and partial differential equations. Laplace transform, theorems of operational calculus, solution of equations with constant coefficients, evaluation of integrals, derivation of transforms, of various functions, etc. For physics, engineering students. Formerly *Modern Operational Calculus.* xiv + 218pp.
60192-7 Paperbound $2.50

PARTIAL DIFFERENTIAL EQUATIONS OF MATHEMATICAL PHYSICS, Arthur G. Webster. Introduction to basic method and theory of partial differential equations, with full treatment of their applications to virtually every field. Full, clear chapters on Fourier series, integral and elliptic equations, spherical, cylindrical and ellipsoidal harmonics, Cauchy's method, boundary problems, method of Riemann-Volterra, many other basic topics. Edited by Samuel J. Plimpton. 97 figures. vii + 446pp. 60263-X Paperbound $2.75

PRINCIPLES OF STELLAR DYNAMICS, Subrahmanyan Chandrasekhar. Theory of stellar dynamics as a branch of classical dynamics; stellar encounter in terms of 2-body problem, Liouville's theorem and equations of continuity. Also two additional papers. 50 illustrations. x + 313pp. 5⅝ x 8⅜.
60659-7 Paperbound $3.00

CELESTIAL OBJECTS FOR COMMON TELESCOPES, T. W. Webb. The most used book in amateur astronomy: inestimable aid for locating and identifying hundreds of celestial objects. Volume 1 covers operation of telescope, telescope photography, precise information on sun, moon, planets, asteroids, meteor swarms, etc.; Volume 2, stars, constellations, double stars, clusters, variables, nebulae, etc. Nearly 4,000 objects noted. New edition edited, updated by Margaret W. Mayall. 77 illustrations. Total of xxxix + 606pp.
20917-2, 20918-0 Two volumes, Paperbound $5.00

A SHORT HISTORY OF ASTRONOMY, Arthur Berry. Earliest times through the 19th century. Individual chapters on Copernicus, Tycho Brahe, Galileo, Kepler, Newton, etc. Non-technical, but precise, thorough, and as useful to specialist as layman. 104 illustrations, 9 portraits, xxxi + 440 pp. 20210-0 Paperbound $3.00

ORDINARY DIFFERENTIAL EQUATIONS, Edward L. Ince. Explains and analyzes theory of ordinary differential equations in real and complex domains: elementary methods of integration, existence and nature of solutions, continuous transformation groups, linear differential equations, equations of first order, non-linear equations of higher order, oscillation theorems, etc. "Highly recommended," *Electronics Industries*. 18 figures. viii + 558pp. 60349-0 Paperbound $3.50

DICTIONARY OF CONFORMAL REPRESENTATIONS, H. Kober. Laplace's equation in two dimensions for many boundary conditions; scores of geometric forms and transformations for electrical engineers, Joukowski aerofoil for aerodynamists, Schwarz-Christoffel transformations, transcendental functions, etc. Twin diagrams for most transformations. 447 diagrams. xvi + 208pp. 6⅛ x 9¼.
60160-9 Paperbound $2.50

ALMOST PERIODIC FUNCTIONS, A. S. Besicovitch. Thorough summary of Bohr's theory of almost periodic functions citing new shorter proofs, extending the theory, and describing contributions of Wiener, Weyl, de la Vallée, Poussin, Stepanoff, Bochner and the author. xiii + 180pp. 60018-1 Paperbound $1.75

AN INTRODUCTION TO THE STUDY OF STELLAR STRUCTURE, S. Chandrasekhar. A rigorous examination, using both classical and modern mathematical methods, of the relationship between loss of energy, the mass, and the radius of stars in a steady state. 38 figures. 509pp. 60413-6 Paperbound $3.25

INTRODUCTION TO THE THEORY OF GROUP'S OF FINITE ORDER, Robert D. Carmichael. Progresses in easy steps from sets, groups, permutations, isomorphism through the important types of groups. No higher mathematics is necessary. 783 exercises and problems. xiv + 447pp. 60300-8 Paperbound $3.50

THE SOLUBILITY OF NONELECTROLYTES, Joel H. Hildebrand and Robert L. Scott. Classic, pioneering work discusses in detail ideal and nonideal solutions, intermolecular forces, structure of liquids, athermal mixing, hydrogen bonding, equations describing mixtures of gases, high polymer solutions, surface phenomena, etc. Originally published in the American Chemical Society Monograph series. New authors' preface and new paper (1964). 148 figures, 88 tables. xiv + 488pp. 61125-6 Paperbound $3.00

INTRODUCTION TO APPLIED MATHEMATICS, Francis D. Murnaghan. Introduction to advanced mathematical techniques—vector and matrix analysis, partial differential equations, integral equations, Laplace transform theory, Fourier series, boundary-value problems, etc.—particularly useful to physicists and engineers. 41 figures. ix + 389pp. 61042-X Paperbound $2.25

ELEMENTARY MATHEMATICS FROM AN ADVANCED STANDPOINT: VOLUME I—ARITHMETIC, ALGEBRA, ANALYSIS, Felix Klein. Second-level approach, illuminated by graphical and geometrical interpretation. Covers natural and complex numbers, real equations with real unknowns, equations in the field of complex quantities, logarithmic and exponential functions, goniometric functions, infinitesimal calculus, transcendence of e and π. Concept of function introduced immediately. Translated by E. R. Hedrick and C. A. Noble. 125 figures. ix + 274pp. (USO) 60150-1 Paperbound $2.25

ELEMENTARY MATHEMATICS FROM AN ADVANCED STANDPOINT: VOLUME II—GEOMETRY, Feliex Klein. Using analytical formulas, Klein clarifies the precise formulation of geometric facts in chapters on manifolds, geometric and higher point transformations, foundations. "Nothing comparable," *Mathematics Teacher.* Translated by E. R. Hedrick and C. A. Noble. 141 figures. ix + 214pp. (USO) 60151-X Paperbound $2.25

ENGINEERING MATHEMATICS, Kenneth S. Miller. Most useful mathematical techniques for graduate students in engineering, physics, covering linear differential equations, series, random functions, integrals, Fourier series, Laplace transform, network theory, etc. "Sound and teachable," Science. 89 figures. xii + 417pp. 6 x 8½. 61121-3 Paperbound $3.00